Edited by
Johan Dubbeldam,
Kirk Green, and Daan Lenstra

The Complexity of Dynamical Systems

Related Titles

Schuster, H. G. (ed.)

Reviews of Nonlinear Dynamics and Complexity

Volume 1

2008

ISBN: 978-3-527-40729-3

Schuster, H. G. (ed.)

Reviews of Nonlinear Dynamics and Complexity

Volume 2

2009

ISBN: 978-3-527-40850-4

Schuster, H. G. (ed.)

Reviews of Nonlinear Dynamics and Complexity

Volume 3

2010

ISBN: 978-3-527-40945-7

Pastor, J.

Mathematical Ecology of Populations and Ecosystems

2008

ISBN: 978-1-4051-8811-1

Vedmedenko, E.

Competing Interactions and Patterns in Nanoworld

2007

ISBN: 978-3-527-40484-1

Mazenko, G. F.

Nonequilibrium Statistical Mechanics

2006

ISBN: 978-3-527-40648-7

Kane, D., Shore, A. (eds.)

Unlocking Dynamical Diversity

Optical Feedback Effects on Semiconductor Lasers

2005

Online Book Wiley Interscience

ISBN: 978-0-470-85621-5

Schuster, H. G., Just, W.

Deterministic Chaos

An Introduction

2005

ISBN: 978-3-527-40415-5

Hinrichsen, H., Wolf, D. E. (eds.)

The Physics of Granular Media

2004

ISBN: 978-3-527-40373-8

Edited by
Johan Dubbeldam, Kirk Green, and Daan Lenstra

The Complexity of Dynamical Systems

A Multi-disciplinary Perspective

WILEY-VCH Verlag GmbH & Co. KGaA

The Editors

Dr. Johan Dubbeldam
Delft University of Technology
Delft Institute of Applied Mathematics
Delft, The Netherlands
J.L.A.Dubbeldam@ewi.tudelft.nl

Dr. Kirk Green
TNO Bouw en Ondergrond
Delft, The Netherlands
kirk.green@tno.nl

Prof. Daan Lenstra
Delft University of Technology
EEMCS
Delft, The Netherlands
d.lenstra@tudelft.nl

All books published by **Wiley-VCH** are carefully produced. Nevertheless, authors, editors, and publisher do not warrant the information contained in these books, including this book, to be free of errors. Readers are advised to keep in mind that statements, data, illustrations, procedural details or other items may inadvertently be inaccurate.

Library of Congress Card No.: applied for

British Library Cataloguing-in-Publication Data
A catalogue record for this book is available from the British Library.

Bibliographic information published by the Deutsche Nationalbibliothek
The Deutsche Nationalbibliothek lists this publication in the Deutsche Nationalbibliografie; detailed bibliographic data are available on the Internet at http://dnb.d-nb.de.

© 2011 WILEY-VCH Verlag & Co. KGaA, Boschstr.12, 69469 Weinheim, Germany

All rights reserved (including those of translation into other languages). No part of this book may be reproduced in any form – by photoprinting, microfilm, or any other means – nor transmitted or translated into a machine language without written permission from the publishers. Registered names, trademarks, etc. used in this book, even when not specifically marked as such, are not to be considered unprotected by law.

Composition Thomson Digital, Noida, India
Printing and Binding Fabulous Printers Pte Ltd, Singapore
Cover Design Schulz Grafik-Design, Fußgönheim

Printed in Singapore
Printed on acid-free paper

ISBN: 978-3-527-40931-0

Contents

Acknowledgements *X*
List of Contributors *XI*

Introduction *1*

Part One Applications *5*

1 Coastal Morphodynamics *7*
Nicholas Dodd
1.1 1D Theoretical Framework *10*
1.2 Sediment Transport, Erosion, and Deposition *11*
1.3 Wave Speeds, Free and Forced Modes, and Timescales *13*
1.4 Phase-Averaged Morphodynamics: Crescentic Bars *17*
1.5 Physical Mechanisms *21*
1.6 Concluding Remarks *24*
References *25*

2 Long-Lived Transients in Transitional Pipe Flow *27*
Jerry Westerweel and Dirk Jan Kuik
References *36*

3 Dynamics of Patterns in Lasers with Delayed Feedback *37*
Kirk Green and Bernd Krauskopf
3.1 Introduction *37*
3.2 Single-Mode Laser with COF *40*
3.3 VCSEL with Optical Feedback *42*
3.3.1 EEM-Reduced Model of a Two-Mode VCSEL with Optical Feedback *44*
3.3.2 Analytical Results for the Two-Mode ECM Structure *45*
3.4 Numerical Bifurcation Analysis of the Two-Mode ECM Structure *47*

3.4.1	Dependence of the Bifurcation Diagram on η 48
3.4.2	Transitions Involving Codimension-two Points Near $\eta = 0.0$ 51
3.4.3	Details of the Transition to $\eta = 1.0$ 53
3.5	Stability and Bifurcations of Periodic Solutions 56
3.6	Discussion 59
	References 60

4 Optical Delay Dynamics and Its Applications 63
Laurent Larger and Ingo Fischer

4.1	Introduction 63
4.2	Experimental Setups 65
4.2.1	Delayed Feedback and Delay-Coupled Semiconductor Lasers 65
4.2.2	Nonlinear Delayed Electro-Optic Feedback 66
4.3	Dynamics: Modeling, Numerics, and Experiments 68
4.3.1	The All-Optical Feedback System and the Lang–Kobayashi Rate Equations 68
4.3.2	The All-Optically Delay-Coupled Lasers 69
4.3.2.1	Two Mutually Delay-Coupled Lasers: Generalized Synchronization and Symmetry Breaking 69
4.3.2.2	Delay-Coupled Lasers with a Relay: Zero-Lag Synchronization 73
4.3.2.3	Rings of Unidirectionally Coupled Lasers with Delay 78
4.3.3	Linear Filtering of a Nonlinear Delayed Feedback 81
4.3.3.1	Modeling 81
4.3.3.2	Delay Differential Dynamics 82
4.3.3.3	Integro-Differential Delay Dynamics 83
4.4	Applications 89
4.4.1	Chaos Communications 90
4.4.2	Spatial or Spectral Stabilization in Semiconductor Lasers 92
4.4.2.1	High Spectral Purity Microwave Oscillation 92
4.4.3	Further Applications of All-Optical Delay Systems 95
4.5	Conclusion and Outlook 95
	References 96

5 Symbolic Dynamics in Genetic Oscillation Patterns 99
Simone Pigolotti, Sandeep Krishna, and Mogens H. Jensen

5.1	Introduction 99
5.2	The Method 100
5.2.1	Monotonicity and Nullclines 100
5.3	The Negative Feedback Loop 101
5.3.1	NFL and Reverse Engineering 105
5.4	Multiple Loops 109
5.4.1	Multiple Loops and Reverse Engineering 113
5.5	Negative Feedback Loops on a Lattice 114
5.6	Conclusions 116
	References 116

6	**Translocation Dynamics and Randomness** *119*	
	Johan Dubbeldam, Vakhtang Rostiashvili, Andrey Milchev, and Thomas Vilgis	
6.1	Introduction *119*	
6.2	Anomalous Diffusion Model *121*	
6.2.1	Unbiased Translocation *122*	
6.2.1.1	Continuous Time Random Walks and Fractional Derivatives *124*	
6.2.2	Monte Carlo Simulations *127*	
6.2.3	Driven Translocation *130*	
6.3	Statistical Moments $\langle s \rangle$ and $\langle s^2 \rangle$ versus Time *132*	
6.3.1	Translocation: A Quasi-Equilibrium Process? *135*	
6.4	Discussion *137*	
	References *137*	

Part Two Fundamental Aspects *139*

7	**Entropy Production, the Breaking of Detailed Balance, and the Arrow of Time** *141*	
	Christian Van den Broeck	
7.1	Introduction *141*	
7.2	Detailed Balance *142*	
7.3	Stochastic Thermodynamics *144*	
7.4	Microscopic Expression of Entropy Production *150*	
7.5	Discussion *154*	
	References *156*	

8	**Monodromy and Complexity of Quantum Systems** *159*	
	Boris Zhilinskii	
8.1	Introduction *159*	
8.2	Hamiltonian Monodromy *161*	
8.3	Classical–Quantum Correspondence *164*	
8.4	Lattices and Defects *166*	
8.4.1	Elementary Monodromy Defects *166*	
8.4.2	Fractional Monodromy Defects *168*	
8.4.3	Monodromy–Defect Correspondence *170*	
8.5	Multicomponent Energy–Momentum Map, Bidromy, and Others *173*	
8.6	Time Evolution and Monodromy *174*	
8.7	Perspectives *175*	
	References *179*	

9	**Dynamics in Materials Science** *183*	
	Gérard A. Maugin and Martine Rousseau	
9.1	Introduction *183*	
9.2	Essentials of Elasticity *185*	
9.2.1	Finite Strain Elasticity *185*	

9.2.2	Small Strain Elasticity	*187*
9.2.3	One-Dimensional Motion	*188*
9.2.4	Physical Nonlinearity	*189*
9.3	The Boussinesq Paradigm and Akin Nonlinear Dispersive Systems	*190*
9.3.1	The Standard Boussinesq (BO) Equation in Elastic Crystals	*190*
9.3.2	The "Good" Boussinesq Equation	*194*
9.3.3	Generalized Boussinesq Equation	*195*
9.3.4	Mechanical System with Two Degrees of Freedom	*196*
9.3.5	Sine-Gordon Equations and Associated Systems of Equations	*197*
9.3.5.1	Pure Sine-Gordon System	*197*
9.3.5.2	Sine-Gordon–d'Alembert Systems	*199*
9.4	A Basic Problem of Materials Science: Phase Transition Front Propagation	*200*
9.4.1	Some General Words	*200*
9.4.2	Microscopic Condensed Matter Physics Approach: Solitonics	*201*
9.4.3	Macroscopic Engineering Thermodynamic Approach	*202*
9.4.4	Mesoscopic Applied Mathematics Approach: Structured Front	*204*
9.4.5	Theoretical Physics Approach: Quasi-Particle and Transient Motion	*205*
9.4.6	Remark on a Mechanobiological Problem	*206*
9.5	Dynamic Materials	*207*
9.6	Further Extensions: Propagation in Metamaterials and Others	*209*
	References	*210*

10 Synchronization on the Circle *213*
Alain Sarlette and Rodolphe Sepulchre

10.1	Introduction	*213*
10.2	Consensus Algorithms on Vector Spaces	*215*
10.3	Consensus Algorithms on the Circle	*217*
10.3.1	Discrete Time	*218*
10.3.1.1	Vicsek Model	*219*
10.3.2	Continuous Time	*219*
10.3.2.1	Kuramoto Model	*220*
10.4	Convergence Properties	*220*
10.4.1	Local Synchronization Like for Vector Spaces	*220*
10.4.2	Some Graphs Ensure (Almost) Global Synchronization	*221*
10.5	Obstacles to Global Synchronization	*222*
10.5.1	Convergence to Local Equilibria for Fixed Undirected $\mathbb{G}$	*222*
10.5.1.1	Hopfield Network	*224*
10.5.1.2	Local Equilibria for the Undirected Ring	*224*

10.5.1.3	Stable Configurations Are Graph Dependent *225*	
10.5.1.4	Structurally Stable Divergent Behavior in Vicsek Model *226*	
10.5.2	Limit Sets Different from Equilibrium *226*	
10.6	Algorithms for Global Synchronization *229*	
10.6.1	Modified Coupling for Fixed Undirected Graphs *229*	
10.6.2	Introducing Randomness in Link Selection *231*	
10.6.2.1	Gossip Algorithm (Directed) *232*	
10.6.3	Algorithms Using Auxiliary Variables *233*	
10.7	Generalizations on Compact Homogeneous Manifolds *234*	
10.8	Conclusions *235*	
	References *238*	

Conclusion and Outlook *241*

Index *243*

Acknowledgements

First of all, we would like to thank all the authors for their excellent contributions to this book and for the generous way in which they dealt with our comments and criticims. We have learned enormously from going through this process with them. Many thanks also to Kees Lemmens of the Applied Mathematics department in Delft, for helping JLAD out with file handling and computer problems. Finally, we express our gratitude to the people of Wiley who made publication of the present work possible. We are especially grateful to Valerie Moliere for her support and encouragement during this project and Ulrike Werner for her help during the later stages in the writing of the present book.

October 2010
Delft University of Technology

Johan L.A. Dubbeldam
Daan Lenstra
Kirk Green

List of Contributors

Nicholas Dodd
University of Nottingham
Environmental Fluid Mechanics
Research Centre
Process and Environmental Division
Faculty of Engineering
University Park
Nottingham NG7 2RD
UK

Johan Dubbeldam
Delft University of Technology
Delft Institute of Applied Mathematics
(DIAM)
Mekelweg 4
2628 CD Delft
The Netherlands

Ingo Fischer
IFISC (UIB-CSIC)
Institute for Cross-Disciplinary Physics
and Complex Systems
Campus Universitat de les Illes Balears
07122 Palma de Mallorca
Spain

Kirk Green
Centrum voor Mechanische en
Maritieme Constructies (CMC)
TNO Bouw en Ondergrond
Van Mourik Broekmanweg 6
2628 XE Delft
The Netherlands

Mogens H. Jensen
The Niels Bohr International Academy
and The Niels Bohr Institute
Blegdamsvej 17
2100 Copenhagen
Denmark

Bernd Krauskopf
University of Bristol
Department of Engineering
Mathematics
Queen's Building
Bristol BS8 1TR
UK

Sandeep Krishna
The Niels Bohr International Academy
and The Niels Bohr Institute
Blegdamsvej 17
2100 Copenhagen
Denmark

Dirk Jan Kuik
Delft University of Technology
Laboratory for Aero & Hydrodynamics
Mekelweg 2
2628 CD Delft
The Netherlands

The Complexity of Dynamical Systems. Edited by J. Dubbeldam, K. Green, and D. Lenstra
Copyright © 2011 WILEY-VCH Verlag GmbH & Co. KGaA, Weinheim
ISBN: 978-3-527-40931-0

Laurent Larger
Université de Franche-Courté
Institut Universitaire de France
Institut FEMTO-ST/Optics,
UMR CNRS 6174
16 route de Gray
25030 Besançon
France

Gérard A. Maugin
Université Pierre et Marie Curie, Paris
Institut Jean Le Rond d'Alembert
UMR CNRS 7190
Case 162, 4 place Jussieu
75252 Paris Cedex 05
France

Andrey Milchev
Bulgarian Academy of Sciences
Institute of Physical Chemistry
Acad. G. Bonchev Str., bl. 11
1113 Sofia
Bulgaria

Simone Pigolotti
The Niels Bohr International Academy
and The Niels Bohr Institute
Blegdamsvej 17
2100 Copenhagen
Denmark

Vakhtang G. Rostiashvili
Max-Planck Institute for
Polymer Research
Ackermannweg 10
55128 Mainz
Germany

Martine Rousseau
Université Pierre et Marie Curie, Paris
Institut Jean Le Rond d'Alembert
UMR CNRS 7190
Case 162, 4 place Jussieu
75252 Paris Cedex 05
France

Alain Sarlette
Université de Liège
Department of Electrical Engineering
and Computer Science
Grande Traverse, 10
4000 Liège Sart-Tilman
Belgium

Rodolphe Sepulchre
Department of Electrical Engineering
and Computer Science
Université de Liège
Grande Traverse, 10
4000 Liège Sart-Tilman
Belgium

Christian Van den Broeck
University of Hasselt
Campus Diepenbeek
Agoralaan building D
3590 Diepenbeek
Belgium

Thomas Vilgis
Max-Planck Institute for
Polymer Research
Ackermannweg 10
55128 Mainz
Germany

Jerry Westerweel
Delft University of Technology
Laboratory for Aero & Hydrodynamics
Mekelweg 2
2628 CD Delft
The Netherlands

Boris Zhilinskii
Université du Littoral
189A, Avenue Maurice Schumann
MREI 2
59140 Dunkerque
France

Introduction

When the study of dynamical systems started some 50 years ago, most attention was paid to systems containing a few degrees of freedom and to the study of the deterministic dynamics of these systems through analytical methods and numerical computations. The aim was to determine the stability boundaries of the system. The mathematical methods to analyze nonlinear systems have developed, since then.

A famous example is the Lorentz equations containing only three degrees of freedom. In particular, this system was shown to exhibit chaos, a phenomenon that has received a lot of attention. It was found that chaos is inherent to a lot of physical models used to describe fluid flows, lasers, cardiac rhythms, and so on. The corresponding sensitivity to initial conditions entailed that small fluctuations caused by noise sources in the system could have dramatic effects on the system trajectories.

Meanwhile, in the physics community, people dealing with fluctuations were mainly working in statistical physics, where the linear response theory of Kubo was one of the great successes. In statistical physics, the approach was certainly much different from that used in dynamical systems theory in that one started from systems with *many* degrees of freedom and the dynamical properties of such systems were then studied by applying small perturbations around the stationary state of the physical system, instead of the arbitrary perturbations, allowed in a bifurcation analysis.

More recently, dynamical systems analysis has grown to encompass many fields of research. The most striking development took place in systems with many degrees of freedom. Unlike in the past, these systems can now be subjected to closer examination thanks to developments in bifurcation theory together with improvements in the numerical packages, which enable one to follow bifurcations in large parameter spaces. This has greatly extended the possibilities of using bifurcation analysis, which has nowadays found application in numerous physical systems, for example, fluid flows modeled by Navier–Stokes equations, (optical) systems with delayed feedback, and biological systems with a large number of components. Through all these developments, the dynamical systems theory has matured and has started to accommodate interdisciplinary aspects.

In the statistical physics community, the attention has gradually moved into the direction of topological properties and symmetries (such as time invariance) in systems with a great number of degrees of freedom. This is demonstrated by huge number of publications that have appeared on networks and various aspects of network growth and interaction among networks, such as synchronization.

The changes that have taken place for the statistical physics community and for the people who have traditionally worked in dynamical systems have led to a common interest in a thriving research field, in which research methods from both fields are successfully combined. This merger of the statistical physics and dynamical systems research has led to the field of *complexity*, an archetypal example of *interdisciplinary research*.

Such a synthesis was clearly reflected in the 2008 Dynamics Days Conference held in Delft, and we strongly felt the need for a book in which these common aspects of dynamical systems research were emphasized. In the spirit of the Dynamics Days Conference, this book consists of a variety of topics, presented in 10 chapters, which were chosen from invited and plenary talks given at the conference. In addition, the book was augmented with some chapters that address additional topics, such as symbolic dynamics in genetics and turbulence, as they contributed well to the spirit of the book, and in these chapters some of the latest developments in specific branches in complexity research are presented. We believe that this book is timely covering a great deal of topics that are receiving a lot of attention from scientists with different backgrounds, ranging from biology, engineering, to physics and applied mathematics.

This volume has two parts. In the first part (Chapter 1–6) several applications of dynamical systems are discussed. The second part of the book is devoted to more fundamental aspects of complexity in dynamical systems. We start with interesting problems that can be tackled with the applied mathematics approach to dynamical systems. Dynamical processes that underlie coast formation form the subject of Chapter 1 by N. Dodd. In Chapter 2, new insights into turbulent pipe flows are presented by J. Westerweel and N. Kuiken. In Chapter 3, K. Green and B. Krauskopf treat optical systems with delayed feedback. This chapter illustrates very well how spatial bifurcation theory has evolved in recent years. Next in Chapter 4, I. Fischer and L. Larger deal with the problem of a great number of delay-coupled laser systems with feedback. As both delay and a great number of systems are taken into account, this is a good example of a problem that requires an interdisciplinary approach. In Chapter 5, M. Jensen and S. Pigolotti discuss symbolic dynamics in gene expression. This is a very nice example of the use of dynamical systems theory in biological problems. In the last chapter of the first part of the book, J. Dubbeldam, V. Rostiashvili, and A. Milchev discuss another application of dynamical systems to biology: translocation of a polymer chain through a nanopore. However, in this case noise and statistical physics also play an essential role.

In Chapter 7, in the second part of the book, entropy production is discussed by C. van den Broeck. This is a nice example of dynamics involved in a statistical physics problem. Moreover, this chapter is connected with Chapter 8 by B. Zhilinsky who discusses monodromy in quantum systems and shows how topological defects and symmetries are related to monodromy. In Chapter 9, G. Maugin and M. Rousseau explain the role of dynamics in elasticity theory. Especially (non)linear wave phenomena and solitons are discussed. The book ends with a chapter by R. Sepulchre and A. Sarlette, dealing with synchronization on a circle. This chapter is a nice example of how the dynamical phenomenon of synchronization depends on the topology of the underlying space. It treats, from a network perspective, a class of coupled system problems, which are similar in nature to those discussed in Chapter 4, for a system of coupled lasers with feedback.

Part One
Applications

1
Coastal Morphodynamics

Nicholas Dodd

As it is related to coastal morphodynamics, the coastal region is, loosely, the region that runs from the interface between sea and land out to the edge of the continental shelf. Coastal morphodynamics is then generally defined as the dynamical interaction of hydrodynamics (waves and currents) with the erodible bed in this region. The kind of interaction that takes place is in a large part dictated both by the type of erodible sediment (cohesive or noncohesive, low or high permeability, and density and shape of sediment) and by the type of hydrodynamics that dominates that part of the coastal region. For instance, in most of the continental shelf it is tidal currents that are primarily "felt" by the seabed, with only storm-generated wind waves or, possibly, long period swell (i.e., storm-generated waves that have propagated some distance from their source, so that they are no longer directly associated with the storm) being felt on occasions.[1] Where surface gravity waves (often called, simply, wind waves due to their most common generating mechanism, as opposed to their restoring force) start to "feel" the bed, they change shape and slow down (the celerity, or phase speed decreases). This region is termed the shoaling zone, and from here on in the seabed is, in turn, affected by the wave motion. Waves then break in the so-called surf zone, when the depth reduces to be of the same order as the height of the waves, and the breaking generates currents both directly along (the long shore current) and, indirectly, offshore (the rip current). Finally, at the interface between sea and land, waves (in the so-called swash zone) and tide (the intertidal zone) both periodically cover the beach or shoreface. In these regions, waves are almost always important in moving sediment and shaping the seabed, but tidal currents can also be of influence when the range is substantial.

For engineering purposes, the important thing to know is whether erosion or deposition will take place and if so where and how rapidly this will occur. The problems of, for instance, scour beneath a pipeline or at a piling are well known. In problems such as these, human interference causes a change in local hydrodynamics, which then feeds back on the stability of the engineering work. Perhaps

---

[1] An exception to this is tsunamis, the motion of which is felt over the whole water column even in the deep ocean. Tsunami morphodynamics will not be discussed here, although the swash morphodynamics that is discussed has direct relevance to that of tsunamis.

The Complexity of Dynamical Systems. Edited by J. Dubbeldam, K. Green, and D. Lenstra
Copyright © 2011 WILEY-VCH Verlag GmbH & Co. KGaA, Weinheim
ISBN: 978-3-527-40931-0

less well known are the kinds of problems that occur due to the preexisting morphodynamics in the coastal region. These kinds of problems occur quite frequently throughout the coastal region. For example, gas and oil pipelines on the continental shelf frequently must traverse sand wave fields, which means that pipes must be laid on top of or be buried (or be completely buried but at different depths) beneath alternate sections of the seabed if large stresses on the pipes are to be avoided. If the sand waves actively migrate (which they do), the problems are severely exacerbated. Therefore, migration rates, amplitudes, and wavelengths of these features are important quantities to be able to estimate (see Figure 1.1). In Figure 1.2, a close-up of a seabed morphology where a channel has been dredged in a field of sand waves can be seen. Note the approximate wavelength and amplitudes of these waves.

These kinds of quasi-rhythmic patterns can be found throughout the coastal region on many different scales. In Figure 1.3, we can see ripple patterns on a beach, which are highly reminiscent of the measured morphology shown on its right, which is actually of sand waves and therefore at a vastly different scale.

In Table 1.1, we see examples of many of the morphological features that can be found in the coastal region [2].

Despite the differences in scales (both spatial and temporal), and in geographical location, these features are all highly organized in that they possess readily identifiable, quasi-regular wavelengths and amplitudes. They also share a common (morpho) dynamics in that they occur not because some hydrodynamic motion (such as a surface gravity wind wave or a Kelvin wave) erodes a pattern in the sandy beach or seabed, but rather as a result of the so-called self-organization. What this means is that the pattern is a natural state that the seabed will evolve into as long as the waves or currents mentioned in Table 1.1 exist.

Figure 1.1 Depiction of the location of sand waves (gray regions) in the southern North Sea. From Ref. [1], copyright (2001) American Geophysical Union.

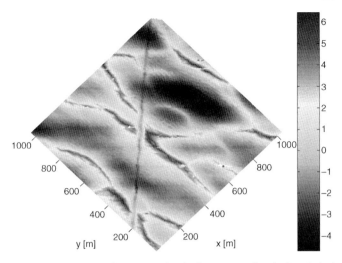

Figure 1.2 Close-up of a section of seabed near Rotterdam harbor. A dredged channel runs approximately vertically through the region. Courtesy of Michiel Knaapen, HR Wallingford Ltd. Color bar in meters.

In this chapter, we present a brief account of some of the fundamental morphodynamics that contribute to their occurrence. To do this, we focus primarily on one type of pattern, namely, crescentic bars.

In the next section, we present a brief theoretical framework and introduce a simple equation system that will serve to illustrate some basic morphodynamics. Thereafter, we look at sediment transport and its representation. We then take a look at hydro- and morphodynamical wave speeds, timescales, and the difference between forced and free modes, to understand the difference that having a mobile (erodible) bed can make to the dynamics.

Figure 1.3 (a) Photograph of ripples on a sandy beach (courtesy of Albert Falqués, Universitat Politecnica de Catalunya). (b) Image of seabed morphology near the entrance to Rotterdam harbor; lighter shading indicates higher bed level (courtesy of Suzanne Hulscher, University of Twente).

Table 1.1 Showing different coastal morphological patterns, the typical necessary conditions for their observation, and their length scales and migration rates.

Feature	Spacing	Height	Migration rate
Tidal currents in a continental shelf sea			
sand banks	5–10 km	5–15 m	—
sand banks	200–700 m	1–10 m	1–10 m/y
Tidal currents along a coast			
shoreface connected ridges	5–8 km	1–5 m	1–10 m/y
Waves on a straight coast			
crescentic and oblique bars	100–500 m	1–5 m	0–20 m/day
transverse bars	10–100 m	1–2 m	—
beach cusps	1–30 m	0.1–1 m	—

1.1
1D Theoretical Framework

In the surf and shoaling zones, at the interface of which crescentic bars can exist, it is common to describe the hydrodynamics using shallow water theory. This amounts to assuming that pressure fields are hydrostatic (so free surface gradients drive the flow[2]) and the flow depth invariant to a first degree of approximation.

In Figure 1.4, the depth-averaged velocity, $u(x, t)$, is shown, which is defined as

$$u(x,t) = \frac{1}{d}\int_{z=z_b}^{z=z_s} u_T(x,z,t)dz,$$

where x and z are horizontal and vertical spatial coordinates (see Figure 1.4), z_s is the position of the instantaneous free surface, z_b is the position of the seabed, d is water depth, and t is time. By using this quantity, a shallow water hydrodynamical system can be derived from the Euler equations of motion (see, for example, [3, 4]):

$$\frac{\partial d}{\partial t} + \frac{\partial (du)}{\partial x} = 0, \qquad (1.1)$$

$$\frac{\partial u}{\partial t} + u\frac{\partial u}{\partial x} = -g\frac{\partial z_s}{\partial x}, \qquad (1.2)$$

where g is gravity.

---

[2] Note that when considering phase-averaged equations, another force, referred to as the radiation stress(es) and akin to Reynolds stresses, arises. We consider this later on.

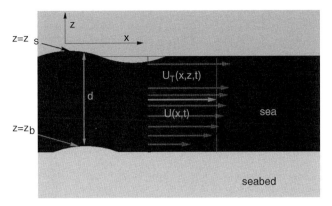

Figure 1.4 Theoretical framework for shallow water hydrodynamics. A schematic depiction of a water/sandy bed system, and the definition of a depth-averaged velocity, $u(x, t)$, from the depth-varying velocity, $u_T(x, z, t)$.

1.2
Sediment Transport, Erosion, and Deposition

Sediment transport is commonly categorized as either bed or suspended load, the former being the movement of grains along the seabed, either by rolling or by jumping (saltating), and the latter being the entrainment of sediment within the water column, which is then transported as part of the fluid flow (see [5, 6]). The so-called total load (q) is the sum of these parts and is written accordingly as

$$q[\text{m}^3/\text{s}/\text{m}] = q_b + q_s, \tag{1.3}$$

where q_b and q_s are bed and suspended load components, respectively, and where the dimensions of q are those of volumetric transport per unit width of seabed, as shown in (1.3).

Although bed load does not travel at the flow speed, it is certainly reasonable to state that $q_b \propto u$. Beyond this simple relation, there are a great number of relations that purport to quantify q_b (see [6]). Physically, there is a threshold of motion below which the drag exerted by the flow on the grains is too small to effect motion, and many formulae do incorporate this effect. Here, we make do with the following simple relation:

$$q_b = Au^m - \gamma \frac{\partial z_b}{\partial x}, \tag{1.4}$$

where, typically, $2 < m < 4$, A is a dimensional constant (the dimensions of which depend on m), and γ is related to the angle of repose of the bed material. The second term here therefore represents the tendency of bed material to move down its own gradient under gravity: that is, bed diffusion. So, the greater the flow speed the greater will be q_b, as the flow moves more and more material along the bed, thus eroding the bed. As the flow decelerates, less material will be moved and so deposition will

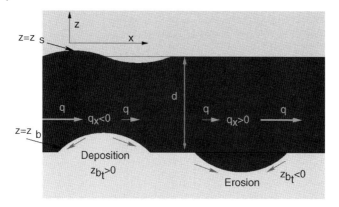

Figure 1.5 Schematic depiction of erosion and deposition by bed load transport. Arrows in the bed indicate the direction of bed load sediment movement due to bed diffusion.

immediately take place. Therefore, the spatial gradient in q_b, q_{b_x} (in 2D the sediment flux divergence, $\vec{\nabla} \cdot \vec{q}_b$) will govern sediment erosion and deposition. This is depicted in Figure 1.5.

In suspended load, the situation is not quite so simple. First, sediment enters the water column, and so the amount present can be represented by a concentration $\tilde{c}$, which will typically vary over the water column: $\tilde{c} = \tilde{c}(x, z, t)$. Thus, q_s is straightforwardly expressed as

$$q_s = \int_{z_b}^{z_s} \tilde{c}(x, z, t) u_T(x, z, t) dz, \tag{1.5}$$

$$q_s \approx d(x, t) C(x, t) u(x, t), \tag{1.6}$$

where C is the depth-averaged concentration (volumetric or mass) per unit area of seabed. Another important difference from bed load is that even if the flow were to cease immediately, the sediment would take a finite time to fall under gravity to the bed (the so-called settling lag). In these circumstances, therefore, sediment transport would cease ($q_s \approx 0$), but much sediment would be (temporarily) stored within the water column (see Figure 1.6).

These dynamics can conveniently be combined into one equation, the sediment conservation or Exner equation (see [5]):

$$\frac{\partial z_b}{\partial t} = -\frac{1}{1-n} \left[\frac{\partial q}{\partial x} + \frac{\partial (dC)}{\partial t} \right], \tag{1.7}$$

where n is porosity.

The inclusion of the second term on the right of (1.7) (and explicit consideration of suspended load and therefore the inclusion of settling lag effects) brings with it significantly more complex dynamics (and greater realism for smaller sediment sizes). For our purposes, however, it suffices to consider only bed load in a formula of type (1.4), which we can further simplify as

$$q_b = Au^m \tag{1.8}$$

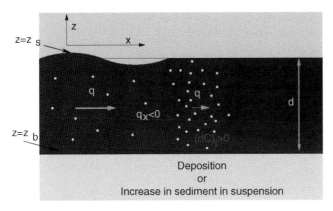

Figure 1.6 Schematic depiction of suspended load transport and the diminution thereof.

We have thus excluded bed diffusion, which will not significantly affect our morphodynamics as its effect is to smooth out (damp) bed waves (see Figure 1.5).

1.3
Wave Speeds, Free and Forced Modes, and Timescales

It is instructive to consider a simple 1D morphodynamic system:

$$\frac{\partial d}{\partial t} + \frac{\partial (du)}{\partial x} = 0, \tag{1.9}$$

$$\frac{\partial u}{\partial t} + u\frac{\partial u}{\partial x} = -g\frac{\partial z_s}{\partial x}, \tag{1.10}$$

$$\frac{\partial z_b}{\partial t} - \frac{1}{1-n}\frac{\partial q}{\partial x} = 0. \tag{1.11}$$

We can write

$$q = \mathcal{A}u \tag{1.12}$$

$$\Rightarrow \mathcal{C} = \frac{\mathcal{A}}{d} \tag{1.13}$$

representing a depth-averaged concentration. Note that $\mathcal{C} \neq C$, as it includes bed load and might be better thought of as a notional quantity derived from q, which is nonetheless useful for our purposes. Our simple system trivially satisfies the simple dynamic equilibrium depicted in Figure 1.7. This is a flat bed and free surface, with a constant flow of discharge $D_0 U_0$ (i.e., $d = D_0$, $u = U_0$).

Now, we impose a small wave propagating at speed c. This must satisfy all equations (1.9–1.11), and so takes the form:

$$z_s(x, t) = z'_{s0} e^{ik(x-ct)}, \tag{1.14}$$

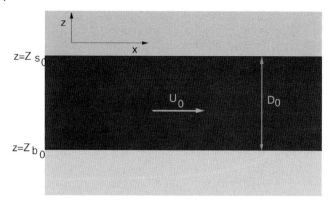

Figure 1.7 Simple 1D morphodynamic system at equilibrium.

$$u(x,t) = U_0 + u'_{s0} e^{ik(x-ct)}, \tag{1.15}$$

$$z_b(x,t) = D_0 + z'_{b0} e^{ik(x-ct)}, \tag{1.16}$$

where the wavelength of the propagating disturbance is $2\pi/k$ and where quantities with primes are small (i.e., $|u'| \ll |U_0|$).

Now, consider for a moment two systems of equations: (i) the hydrodynamic system comprising (1.9) and (1.10) and (ii) the full morphodynamic system comprising (1.9–1.11). In either case, linearization yields

$$c^2 - 2U_0 c + U_0^2 - gD_0 = 0 \quad \Rightarrow \quad c = U_0 \pm \sqrt{gD_0}, \tag{1.17}$$

$$c^3 - 2U_0 c^2 + \left(U_0^2 - gD_0 - \frac{g\mathcal{A}}{(1-n)}\right)c + \frac{g\mathcal{A}}{(1-n)} = 0 \tag{1.18}$$

as the two dispersion relations for (1.17) and (1.18). It can be immediately seen that if $\mathcal{A} = 0$, then (1.18) reduces to (1.17) (see Figure 1.8).

It can be seen that as $\mathcal{A}$ increases and the sediment becomes more mobile, the hydrodynamic (or H_M) wave speeds of the morphodynamic system (i.e., those curves that asymptote to the hydrodynamic (H) wave speeds $c = U_0 \pm \sqrt{gD_0}$ as $\mathcal{A} \to 0$) diverge from their counterparts. Furthermore, the bed (or M) wave speed gradually increases, although it remains far smaller than the H_M wave speeds.

Our simple morphodynamic system (1.9)–(1.11) therefore possesses three so-called free modes, two of which are, as it were, modifications of the H wave speeds and third one being fundamentally new: an M wave. As we shall see later, although in Figure 1.8 it is clear which mode is which, it is in fact not always easy to distinguish between the different modes. Furthermore, remember that each free mode is a solution in all variables. So, for instance, an H_M wave, which will logically be a surface gravity wave propagating close to the classical shallow water wave speed of Airy (see [3]), will also have an expression at the seabed, namely, a wave in the erodible bed traveling at that same speed. We can view this kind of bed wave as being *a forced* wave in that it can be seen as a reaction of the bed to the surface gravity wave. The true M

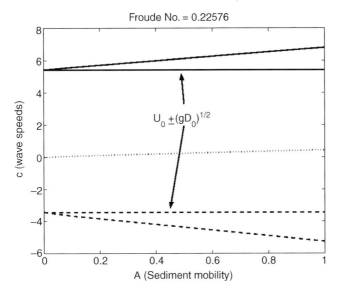

Figure 1.8 Effect of varying sediment mobility parameter $\mathcal{A}$ on the eigenvalues of the simple 1D morphodynamic system at equilibrium. Gray lines indicate the eigenvalues (c) of the system (1.9)–(1.11). Black lines are those of (1.9) and (1.10).

wave, as can be seen from Figure 1.8, will (usually) possess a celerity significantly different from that of the H_M wave, and would not exist without the bed being mobile.

We have seen from Figure 1.8 that the M and H_M wave speeds are generally significantly different. In other words, the hydrodynamic (H) and morphodynamic (M) timescales are different. We can understand this from a more physical point of view by looking at the scalings in the system (1.9)–(1.11). In the shoaling zone, just prior to wave breaking, and for wave heights that are fairly small (perhaps just a gentle swell is present), we can characterize the various length and time (and velocity) scales as

$$d = D_0 \hat{d} : \quad D_0 = 2 \text{ m} \tag{1.19}$$

$$u = U_0 \hat{u} : \quad U_0 = 0.5 \text{ m/s}, \tag{1.20}$$

$$z_s = z_{s0} \hat{z}_s : \quad z_{s0} = 0.5 \text{ m}, \tag{1.21}$$

$$\Delta z_b = z_{b0} \Delta \hat{z}_b : \quad z_{b0} = 0.2 \text{ m}, \tag{1.22}$$

$$t = T_H \hat{t}(\text{in}(1.9), (1.10)) : \quad T_H = 10 \text{ s}, \tag{1.23}$$

$$t = T_M \hat{t}(\text{in}(1.11)), \tag{1.24}$$

$$\mathcal{A} = 0.002 \text{ m}^3/\text{m}^2 \Rightarrow \mathcal{C} = 0.001, \tag{1.25}$$

$$c_H = U_0 + \sqrt{gD_0} = 5 \text{ m/s}, \tag{1.26}$$

$$\Rightarrow L = c_H T_H = 50 \text{ m}, \tag{1.27}$$

$$n = 0.4. \tag{1.28}$$

Note that c_H is the H wave (phase) speed, L is the wavelength of the surface gravity waves, which provides our horizontal, hydrodynamic (H) length scale, L. T_H is the wave period. All the numerical values stated are crude estimates for various physical quantities in this region (or derived therefrom).

If we now nondimensionalize (1.9)–(1.11) using these relations, we get

$$\frac{\Delta D_0}{T_H} \frac{\partial \hat{d}}{\partial \hat{t}} + \frac{\Delta(D_0 U_0)}{L} \frac{\partial (\hat{d}\hat{u})}{\partial \hat{x}} = 0,$$

$$\frac{\Delta U_0}{T_H} \frac{\partial \hat{u}}{\partial \hat{t}} + \frac{U_0 \Delta U_0}{L} \hat{u} \frac{\partial \hat{u}}{\partial \hat{x}} = -\frac{g \Delta z_{s0}}{L} \frac{\partial \hat{z}_s}{\partial \hat{x}}$$

$$\frac{\Delta z_b}{T_M} \frac{\partial \hat{z}_b}{\partial \hat{t}} + \frac{A U_0}{(1-n)L} \frac{\partial \hat{q}}{\partial \hat{x}} = 0$$

and if we now use the above numerical values, as well as characterize changes (Δ) in H variables as those due to the passage of a wave, we find (crudely)

$$0.1 \frac{\partial \hat{d}}{\partial \hat{t}} + 0.04 \frac{\partial \hat{d}\hat{u})}{\partial \hat{x}} = 0,$$

$$0.1 \frac{\partial \hat{u}}{\partial \hat{t}} + 0.01 \hat{u} \frac{\partial \hat{u}}{\partial \hat{x}} = -0.2 \frac{\partial \hat{z}_s}{\partial \hat{x}},$$

$$\frac{0.2}{T_M} \frac{\partial \hat{z}_b}{\partial \hat{t}} + 0.00005 \frac{\partial \hat{q}}{\partial \hat{x}} = 0,$$

where we have omitted units. We can now start to see the importance of the differences between hydrodynamic (T_H) and morphodynamic (T_M) timescales. If we are interested in bed changes at the scale of T_H (i.e., one wave period), $\Rightarrow T_M = T_H$, then the time derivative of the sediment conservation Equation 1.11 is much larger than the divergence term, meaning that there is no effective change in the bed, as might be expected on this sort of timescale. In contrast, if we choose T_M such that the two terms of (1.11) balance, then we find $T_M \approx 1$ h. If we then take $T_H = T_M$, we find that the time derivatives of (1.9) and (1.10) are much smaller than the divergence or advective acceleration/surface gradient terms of those equations. In other words, at the morphodynamic timescale T_M, we can reasonably assume that hydrodynamics adjusts instantaneously, so that the flow is steady (and nondivergent).

Let us now examine a case when this separation in timescales, which does apply in most coastal problems, does not pertain. This happens, for instance, as the water depth becomes very small, such as in the swash. In Figure 1.9, we see the counterpart to Figure 1.8; here, in contrast, depth is varied, and the effect on the wave speeds is marked. The H and H$_M$ wave speeds ($c_H = U_0 \pm \sqrt{gD_0}$) are shown, with a clear transition between sub- and supercritical flow at $U_0/\sqrt{gD_0} = 1$. The M wave speeds,

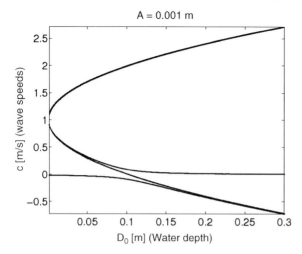

Figure 1.9 Effect of varying depth on the eigenvalues of the simple 1D morphodynamic system at equilibrium. Gray lines indicate the eigenvalues (c) of the system (1.9)–(1.11). Black lines are those of (1.9) and (1.10).

however, show a notably different behavior. For large depths (smaller Froude numbers), the H_M wave speeds are clearly very close to those of the H wave, with the M wave being a small (positive) value. But if we follow the negative H_M wave as depth $\to 0$, then we note that it never becomes positive. Instead, it peels off the corresponding H curve and starts to "look" like an M wave. In contrast, the M wave, which is always > 0, starts behaving very dynamically, acquiring a large positive magnitude, and asymptotes to the negative H wave. In other words, there is an ambiguity in the nature of these waves at large Froude numbers, and waves *can* propagate against the supercritical flow in a fully coupled morphodynamic system (for a shallow water hydrodynamic system, only a shock wave can do this).

An example of a flow structure in the swash, the motion of which in some ways mirrors that shown in Figure 1.9, can be seen in Figure 1.10. Here, the structure on the left, which is reminiscent of a backwash bore and which is propagating onshore (to the left), is shortly to be overwhelmed by the next incoming wave from the right.

1.4
Phase-Averaged Morphodynamics: Crescentic Bars

Crescentic bars are commonly seen in and at the edge of the surf zone on sandy, straight beaches – indeed, it is the waves breaking on them that reveals their presence. Therefore, finite depths usually pertain, Froude numbers are generally small, and so morphodynamic decoupling can be assumed. It is usual not only to use depth-averaged but also phase-averaged equations to describe their dynamics, and this is what here we look at.

Figure 1.10 Photograph of swash backwash on a natural beach. The next wave is approaching from the right. Photograph courtesy of Nicholas Dodd.

In Figure 1.11, the phase-averaging operation is depicted as it relates to waves shoaling in the nearshore region. The depth- and phase-averaged velocity becomes

$$U(x,t) = \frac{1}{T}\int_0^T \frac{1}{d}\int_{z_b}^{z_s} u_T(x,z,t)\,\mathrm{d}z\,\mathrm{d}t = \bar{u}$$

and new variables describing the waves themselves arise

$$E = \frac{1}{8}\varrho g H^2, \quad \text{where} \quad H = \text{wave height}, \tag{1.29}$$

$$\vec{\kappa} = (\kappa_1, \kappa_2), \quad \text{where} \quad \kappa \text{ is the wave number}: \kappa = |\vec{\kappa}|, \tag{1.30}$$

where direction is given by $\tan^{-1}\kappa_2/\kappa_1$ and where $L = 2\pi/\kappa$ (see [3]).

The equations of motion *at the morphodynamic timescale*, which, if we are interested in the bed forms is what we are considering here, become

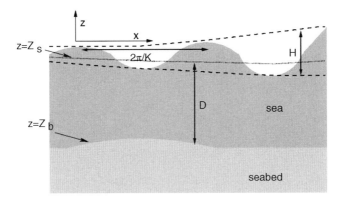

Figure 1.11 Schematized geometry of the nearshore region.

$$\frac{\partial DU_i}{\partial x_i} = 0, \qquad (1.31)$$

$$U_i \frac{\partial U_j}{\partial x_i} = -g\frac{\partial Z_s}{\partial x_j} - \frac{1}{\varrho D}\frac{\partial}{\partial x_i}S'_{ji} + \frac{1}{\varrho D}\frac{\partial}{\partial x_i}S''_{ji} - \frac{\tau_{bj}}{\varrho D}, \qquad (1.32)$$

$$\frac{\partial}{\partial x_i}((U_i + c_{gi})E) + S'_{ij}\frac{\partial U_j}{\partial x_i} = -D, \qquad (1.33)$$

$$\frac{\partial \omega}{\partial x_j} = 0, \qquad (1.34)$$

$$\frac{\partial Z_b}{\partial t} + \frac{1}{1-n}\frac{\partial q_i}{\partial x_i} = 0, \qquad (1.35)$$

where summation is on i and where $D = \bar{d}$, $Z_s = \bar{z}_s$, and $\omega = 2\pi/T_H$, where, it may be recalled, T_H is the wave period. Note that ω is the so-called absolute angular frequency of the wave, namely, the frequency that would be measured by a motionless observer. Furthermore, $\vec{\tau}_b$ is the bed shear stress vector, c_{g_i} are components of the group velocity vector, S'_{ij} are components of the radiation stress tensor, and S''_{ij} are components of turbulent momentum diffusion tensor (see [3, 4] for a more detailed explanation).

In much the same way as we did with (1.9)–(1.11), and in order to determine what kind of solutions there are and whether they are likely to grow, we now seek a time-invariant solution to (1.31)–(1.35). Physically, we may reasonably assume a solution that is uniform along the shore because long straight beaches, if devoid of bed forms, frequently are quasi-uniform alongshore. On this so-called *basic state*, we impose an alongshore periodic disturbance, which can propagate as before

$$Z_s(x, y, t) = Z_{s0}(x) + Z'_s(x)e^{ik(y-ct)}, \qquad (1.36)$$

$$U(x, y, t) = U'(x)e^{ik(y-ct)}, \qquad (1.37)$$

$$V(x, y, t) = V'(x)e^{ik(y-ct)}, \qquad (1.38)$$

$$E(x, y, t) = E_0(x) + E'(x)e^{ik(y-ct)}, \qquad (1.39)$$

$$\kappa_j(x, y, t) = \kappa_{j_0}(x) + \kappa'_j(x)e^{ik(y-ct)}, \qquad (1.40)$$

$$Z_b(x, y, t) = Z_{b0}(x) + Z'_{b0}(x)e^{ik(y-ct)}. \qquad (1.41)$$

Substituting these into (1.31)–(1.35) and linearizing yields a dispersion relation of the form

$$F\vec{\Phi} = \sigma B\vec{\Phi},\qquad(1.42)$$

where

$$\vec{\Phi} = [Z'_s, U', V', E', \kappa'_1, \kappa'_2, Z'_b]^T$$

and

$$\sigma = ck.$$

Note that σ is the so-called intrinsic frequency, which *can* vary in space (see [3]). Now, (1.42) is an ordinary differential equation in x, but if we discretize it we obtain a polynomial of order $6n$, where n is, typically, the number of computational nodes or collocation functions and so on. In general, therefore, c is complex, and as numerical convergence is achieved (as $n \to \infty$), the physical eigenvalues may also be complex. Therefore, some solutions are free modes that may be growing or decaying; $\Rightarrow$ instabilities can exist, implying that there exist some solutions that may grow.

In Figure 1.12, the typical so-called growth rate curves for a dispersion relation such as (1.42) are shown. The growth rate is the imaginary part of σ, and the growth rate curves of Figure 1.12 reveal peaks (i.e., wavelengths or spacings, $2\pi/k$, of morphological patterns that evolve fastest, compared to other local values of k) at two positions in k space. These, as shown in Figure 1.12, correspond to two types of

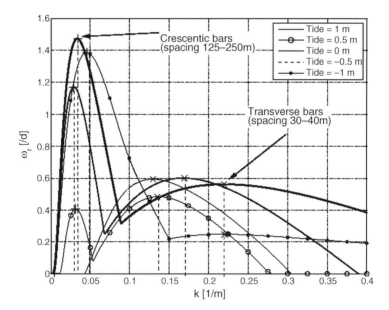

Figure 1.12 Typical growth rate curves for evolving alongshore periodic bed forms on an alongshore uniform coast for various water levels. The thick black line depicts one growth rate curve for crescentic bars. The thick gray line depicts the equivalent growth rate curve for transverse bars. The dashed lines show the point in k space of the fastest growing wavelength (wave number), for each crescentic/transverse bar curve. Courtesy of Meinard Tiessen, University of Nottingham.

evolving bed form: transverse bars, at a smaller spacing, and crescentic bars. In this instance, these are nonmigratory (because waves are normally incident to the beach considered).

In Figure 1.13, we can see the snapshots of the growth to finite amplitude of crescentic bars on a barred beach, and also the wave-generated currents associated with the growing bed forms. Note the simulation starts from a random perturbation to the beach, and the development is independent of that perturbation.

1.5 Physical Mechanisms

We can begin to understand *why* these patterns evolve in the nearshore region by considering the two continuity equations (1.31) and (1.35), which can be combined into one equation (see [7]) such that

$$\frac{\partial Z_b}{\partial t} = -\frac{1}{1-n} \vec{\nabla} \bar{C} \cdot D\vec{U}. \tag{1.43}$$

If we look at Figure 1.14, we can see the mechanism that results in the growth of a perturbation in the seabed in this region. In the first scenario in this figure, a dynamic, time-invariant equilibrium is depicted: $\vec{\Phi}_0 = [Z_{s0}, 0, 0, E_0, \kappa_0, 0, Z_{b0}]^T$. In this equilibrium, waves approach the beach (normally), shoal (i.e., change height and wavelength), primarily according to (1.33) and (1.34), and then break as the water

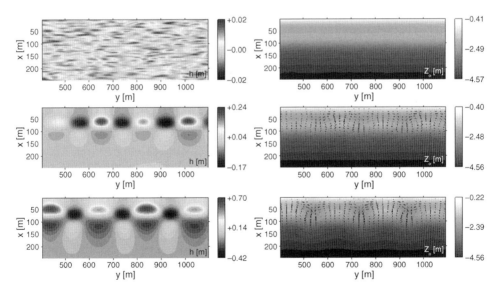

Figure 1.13 Evolving crescentic bars on a barred beach under normal wave incidence. *Top*: initial conditions. *Middle*: after 13 days. *Bottom*: after 200 days. *Left*: bed perturbations. *Right*: currents (arrows) on bed level (light region = shoal; dark region = trough regions). Courtesy of Roland Gamier, University of Nottingham.

depth becomes too small; further shoreward, the wave height remains depth limited. Note that the setup caused by breaking has been ignored in this figure, that is, $Z_s = 0$. The wave breaking induces turbulence in the water column and suspends much sediment, hence the increase in concentration in the vicinity of the breaking point, leading to a positive gradient in concentration shoreward of the breakpoint.

In the second scenario, there is a positive perturbation (i.e., a deposit of sediment) just shoreward of the breakpoint. The presence of this local elevation in the bed (i.e., a shoal) induces the wave height to decay a little more (because the wave heights in the surf zone are, to a first approximation, directly proportional to the water depth, which has been reduced by the presence of the shoal), and this excess decay in wave height induces an onshore current (i.e., $U < 0$ in our coordinate system here) that is generated by the increased local gradient in wave height, represented in the radiation (Reynolds) stress terms S'_{ij} in (1.33) and (1.32).[3] Note then, looking at (1.43), that $-\vec{\nabla}\mathcal{C} \cdot D\vec{U} > 0 \Rightarrow Z_b$ increases. In other words, there is a positive feedback such that a bump (positive bed perturbation) grows further. Similarly, a negative perturbation is accompanied by an offshore current (the rip current), and so Z_b decreases further. In this instance, therefore, the equilibrium $\vec{\Phi}_0$ is unstable, and naturally occurring perturbations in this region will grow.

The rate of growth of these perturbations (bed forms) is proportional to $-\vec{\nabla}\mathcal{C} \cdot D\vec{U}$, so that a strong gradient in $\mathcal{C}$ coupled with a strong current will lead to a relatively rapid growth.

The regularity of these patterns (i.e., the quasi-constant spacing between rip currents) indicates that there is a preferred alongshore length scale, that is, one that grows fastest and therefore dominates other possible length scales, hence the observed regularity. This length scale depends much on the basic geometry of the morphodynamical system. Here, the strong gradients in $\mathcal{C}$ occur at the breakpoint (x_b), which is located where the water depth is sufficiently small. In practice, this often occurs at an alongshore bar (i.e., an elongated shoal running parallel to the shoreline), which can precipitate breaking. In any case, the basic circulation that accompanies growth of crescentic bars is constrained by the presence of the shore on the one hand and of the shoaling zone (where waves do not break, so that the large concentrations at the breakpoint no longer exist) on the other. Typically, this spacing can be between about $2x_b - 3x_b$ [8], although some studies predict larger spacings (e.g., [9]); in the field $4x_b$ is about the mean, although the spread is large (see [10]). The physical reasons of this spacing are not entirely clear. Certainly, a very small spacing will not be favored, not least because bed diffusion would then be large and would significantly damp the positive feedback mechanism (because of larger bed gradients).[4] Why very large spacings might not develop is less obvious, although it seems clear that in some sense the length scale that develops is the most "efficient" in allowing the positive feedback mechanism to operate.

---

3) This basic hydrodynamics has been reproduced in model studies with a fixed bed. See, for example, Ref. [8].
4) Bed diffusion is not present in (1.43), but can be included (see [7]).

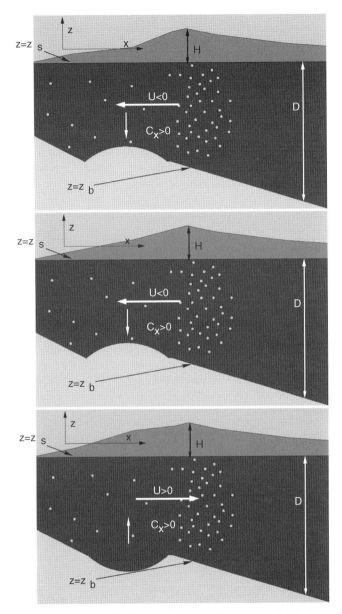

Figure 1.14 Positive feedback mechanism that leads to growth in bed perturbations. *Top*: Dynamic equilibrium at a beach (normal incidence). *Middle*: Positive perturbation: Deposition ⇒ More deposition. *Bottom*: Negative perturbation: Erosion ⇒ More erosion.

1.6
Concluding Remarks

In this brief chapter, we have discussed some of the basics of coastal morphodynamics, with particular reference to one, wave-generated morphological feature: the crescentic bar.

Much has been understood over the past 10–15 years, as the various areas of coastal morphodynamics have been explored. This has been achieved by theoretical, numerical, and field investigations, and each discipline plays a very important part in pushing forward the boundaries of our understanding. Earlier, pioneering investigations have also proved invaluable in posing intriguing questions for later generations of scientists and engineers to investigate with the benefit of increased computing power. Much, however, remains to be understood. Some of these areas for advancement are as follows:

1) **Bed load dynamics**: Our ability to predict fluid motion in view of the known driving forces is good – given enough computing power. However, our understanding of fluid dynamical interaction with soil mechanics and, even more, our ability to predict this are poor. Frequently, a sediment conservation equation alone is used, so that dynamical interactions are ignored. The system is then closed with an empirical sediment transport formula. This approach has led to great advances, but the level of uncertainty, particularly related to quantitative predictions, is very high. (Sediment transport predictions from different formulas that are in agreement within a factor of two are usually considered very close). For suspended load, the situation is not so difficult because the sediment moves with the fluid.

Further fluid dynamical equations are sometimes introduced to describe fluid/sediment regions (typically for sheet flow-type regimes), but the stresses created within the pore regions of the soil by fluid motions, which are related to the initiation of sediment movement, are poorly understood.

2) **Grain dynamics**: Related to the above is the issue of modeling each sediment grain. This is likely to remain prohibitively computationally expensive, although it is conceptually very attractive because the problem at hand then becomes a better defined one of a fluid interacting with solid bodies that are negatively buoyant. Preliminary work looks promising, however, and one value in this approach is likely to be the upscaling and parameterization of larger scale sediment dynamics, so that fluid dynamical-type descriptions can be better formulated.

3) **Improved and/or novel asymptotic and numerical methods**: Even for a simple shallow water, morphodynamical system, analytical solutions are very hard to find, not least because choice of a different sediment transport formula yields a new system. Much progress has been made with weakly nonlinear descriptions, although these are traditionally time consuming to derive and, by their nature, limited in scope. New approaches may yield new insights. Similarly, present numerical methods must cope with a variety of sediment transport formulas, and so must be designed for such.

4) **Sediment movement/entrainment by violent water motions**: How much sediment is put in motion by wave breaking is still a topic of investigation and is very difficult to study either experimentally or by other means. Yet it is crucial for understanding the beach dynamics and, in particular, whether beaches will accrete or erode.

5) **Large-scale sediment/bed form movement**: Residual sediment movement. Most coastal morphodynamics is ultimately related to waves, if tides are also acknowledged to be such, and this draws a clear delineation between fluvial and coastal morphodynamics. As such, residual sediment movement, particularly under surface gravity waves, is the difference between two large quantities; and if we are unable to make predictions with reasonable accuracy of each of the large quantities, then we cannot reasonably hope to accurately capture the small residual. This remains a daunting problem both at a fundamental and at an applied (engineering) level, and is related to all of the above.

References

1 Hulscher, S.J.M.H. and Van den Brink, M. (2001) Comparison between predicted and observed sand waves and sand banks in the North Sea. *J. Geophys. Res.*, **106** (C5), 9327–9338.

2 Dodd, N., Blondeaux, P., Calvete, D., De Swart, H.E., Falqués, A., Hulscher, S.J.M.H., Różyński, G., and Vittori, G. (2003) The use of stability methods for understanding the morphodynamical behavior of coastal systems. *J. Coastal Res.*, **19** (4), 849–865.

3 Mei, C.C. (1990) *The Applied Dynamics of Ocean Surface Waves*, vol. 1, 2nd edn, Advanced Series on Ocean Engineering, World Scientific, Singapore.

4 Svendsen, L.A. (2006) *Introduction to Nearshore Hydrodynamics*, World Scientific.

5 Fredsøe, J. and Deigaard, R. (1993) *Mechanics of Coastal Sediment Transport*, vol. 3, Advanced Series on Ocean Engineering, World Scientific, Singapore.

6 Soulsby, R.L. (1997) *Dynamics of Marine Sands*, Thomas Telford, London.

7 Falqués, A., Coco, G., and Huntley, D.A. (2000) A mechanism for the generation of wave driven rhythmic patterns in the surf zone. *J. Geophys. Res.*, **105** (C10), 24071–24087.

8 Calvete, D., Dodd, N., Falques, A., and Van Leeuwen, S.M. (2005) Morphological development of rip channel systems: normal and near normal wave incidence. *J. Geophys. Res.*, **110** (C10), C10007. doi: 10.1029/2004JC002803

9 Deigaard, R., Drønen, N., Fredsee, J., Jensen, J.H., and Jergensen, M.P. (1999) A morphological stability analysis for a long straight barred beach. *Coastal Eng.*, **36** (3), 171–195.

10 Falqués, A., Dodd, N., Gamier, R., Ribas, F., MacHardy, F., Sancho, L.C., Larroudé, P., and Calvete, D. (2008) Rhythmic surf-zone bars and morphodynamic self-organization. *Coastal Eng.*, **55** (7–8), 622–641. doi: 10.1016/j.coastaleng.2007.11.012

2
Long-Lived Transients in Transitional Pipe Flow
Jerry Westerweel and Dirk Jan Kuik

The transition to turbulence in pipe flow has remained an unsolved problem in fluid mechanics. The transition from laminar pipe flow to a turbulent flow state was first investigated in detail by O. Reynolds in 1883 [1], after whom the Reynolds number is named, defined as $Re = UD/\nu$, where U is the bulk velocity, D the pipe diameter, and ν the kinematic viscosity of the fluid. Typically, for flow rates with a Reynolds number less than 1600 the flow is laminar, while for Reynolds numbers larger than 2000 the flow is strongly intermittent, and laminar and turbulent flow domains coexist [2]. These localized turbulent flow regions are called "puffs." However, a mathematical analysis of the laminar flow state, characterized by a parabolic velocity profile, known as Hagen–Poiseuille (HP) flow, shows that it is linearly stable for all Reynolds numbers [3]. Hence, one is not able to explain the transition to turbulence by means of an instability originating from infinitesimal disturbances, and the transition to turbulence in pipe flow remains unexplained.

A breakthrough occurred when new solutions were found for the flow through a pipe [4, 5]. Each of these solutions, in the form of a *traveling wave* (TW), is an exact solution of the (nonlinear) equations of motion, or Navier–Stokes equations. These TWs are families of solutions characterized by their symmetry. Each TW solution has the character of an unstable saddle, so that one cannot create these solutions under experimental conditions. However, flow patterns that have a very strong reminiscence of these TW solutions could be identified in the experimental data by Hof *et al.* (see Figure 2.1) [6]. The TW solutions first appear for a Reynolds number of about 773 in a mirror-symmetric form. At slightly higher Re, helical and asymmetric TWs are found [7]. At $Re \approx 1300$, those with a twofold and threefold rotational symmetry appear. All these TW solutions have a phase speed that is slightly higher than the mean flow speed (see Figure 2.2).

In order to interpret the observed transition to turbulence in a pipe, we now consider the dynamical behavior of pipe flow in relation to the HP flow and TW flow solutions in terms of a representation in state space. HP flow is then represented as a single stable node. At low Reynolds numbers all disturbances to the base flow decay back to the HP flow, which is represented in state space by a trajectory that returns to the stable node.

The Complexity of Dynamical Systems. Edited by J. Dubbeldam, K. Green, and D. Lenstra
Copyright © 2011 WILEY-VCH Verlag GmbH & Co. KGaA, Weinheim
ISBN: 978-3-527-40931-0

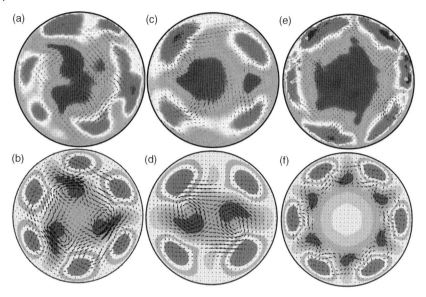

Figure 2.1 Comparison of experimental data of nonstationary pipe flow in a planar cross section (a, c, and e) and corresponding exact traveling wave states (b, d, and f). From Ref. [6].

The TW solutions form a strange repeller. When the HP flow is disturbed sufficiently, the flow state wanders around in phase space, occasionally approaching states that are close to a TW, but each time it is carried away along one of the unstable directions of the unstable saddle, and eventually returns to the laminar flow state (see

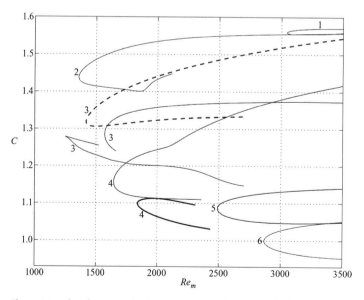

Figure 2.2 The phase speed of traveling waves (the number indicates the rotational symmetry) relative to the mean bulk velocity as a function of the Reynolds number. From Ref. [5].

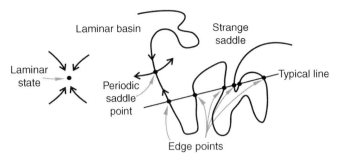

Figure 2.3 Schematic representation of phase space in pipe flow. The laminar flow state is an attractor; by increasing the Reynolds number, its basin of attraction reduces and if the flow is perturbed outside the basin of attraction, the flow wanders around the traveling wave solutions found by Hof *et al.* [4, 5]. Courtesy T. Schneider, University of Marburg.

Figure 2.3). As the Reynolds number increases, the time it takes to return to the stable HP node increases, implying longer and longer transients with increasing Reynolds number. The time the flow state follows a complex trajectory shows large variations. This behavior is typical of a chaotic saddle [8].

To explain a transition to sustained turbulence, it is expected that at a given Reynolds number the strange repeller changes into a strange attractor so that the orbit of a disturbed HP flow no longer returns to the fixed point representing the stable base flow. In that case, HP and turbulent flow coexist, and the two flow states are separated by a boundary that defines the basin of attraction for the laminar and turbulent flow states [9, 10]. For small-amplitude disturbances, the flow quickly returns to the laminar flow state. However, when the disturbance amplitude is large enough, the trajectory passes the boundary for the basin of attraction of the turbulent strange attractor and thus will no longer return to the laminar base flow. This then represents a sustained turbulent flow state that explains the transition to turbulence in a pipe flow.

This transition scenario was investigated by Faisst and Eckhardt [11] by means of a direct numerical simulation (DNS). They simulated the time evolution of localized turbulent flow (representing the turbulent portion of a puff) in a domain five times the pipe diameter with periodic boundary conditions. Although only a small portion of the entire puff is simulated, the essential dynamics of the entire system is captured. Starting at low Re, they determined for each realization how long it took for the flow to return to the laminar flow state. They found that there is a large variation in the time that the flow takes to return to the base state (see Figure 2.4) and that the probability of decay follows an exponential distribution $P \sim \exp(-(t-t_0)/\tau)$, where τ is the characteristic lifetime. The probability distributions for increasing Re are shown in Figure 2.5. This clearly shows that τ increases with Reynolds number.

By taking the time for which 50% of the disturbances had decayed, or median lifetime, they initially found that the lifetime τ diverges at a finite value of the Reynolds number, indicating a critical Reynolds number Re_c of 2250. The value of Re_c was obtained from an *extrapolation* of τ^{-1} as a function of Re.

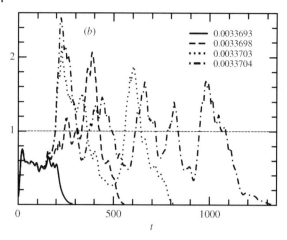

Figure 2.4 Time series of the disturbance energy for four slightly different initial conditions for a pipe flow at $Re = 2000$, showing the large variation of lifetime. From Ref. [11].

Peixinho and Mullin [12] performed an experiment in a pipe flow where they determined the lifetime of puffs. They generated puffs at $Re = 1900$, reduced the flow rate to a lower Re, and then *visually* determined the moment of decay. Like the numerical simulation, the probability showed an exponential decay. From these data, they determined the reciprocal lifetime as a function of Re, and determined from an extrapolation that $Re_c = 1750 \pm 10$, where the lifetime appears to diverge. It should be noted that the uncertainty of the last three to four data points at the highest Re measured essentially occur at the same Re (see Figures 2.6–2.8).

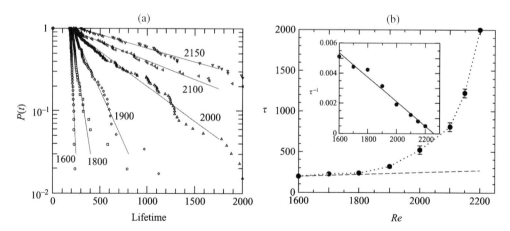

Figure 2.5 Turbulent lifetimes as a function of Reynolds number. (a) Probability $P(t)$ for a single trajectory to be still turbulent after a time t. (b) Median τ of the turbulent lifetimes as a function of Reynolds number. The inset shows the reciprocal median lifetime versus Re and a linear fit, corresponding to $\tau(Re) \propto (Re_c - Re)^{-1}$, with $Re_c \approx 2250$. From Ref. [11].

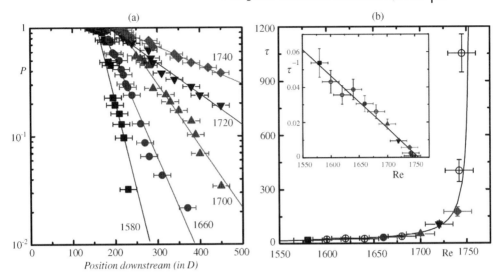

Figure 2.6 (a) Probability of observation of a puff versus downstream distance. (b) Mean decay rate as a function of Re. The inset shows the inverse lifetime with a linear fit indicating $Re_c \approx 1750 \pm 10$. From Ref. [12].

Willis and Kerswell [13] used DNS to simulate the decay of puffs in a pipe at various Reynolds numbers. The DNS was implemented with periodic boundary conditions, with a pipe length 50 times the pipe diameter, which would be long enough to contain the entire puff. The approach was very similar to the experiments performed by Peixinho and Mullin [12]: the simulation was started with a randomly selected snapshot of the velocity field of a puff in the domain at $Re = 1900$ and started at the

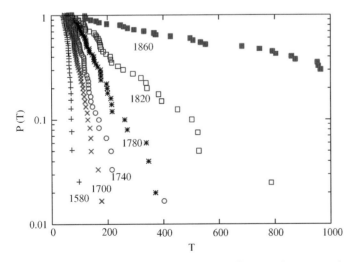

Figure 2.7 Probability $P(T)$ of the lifetime of a puff to exceed T. From Ref. [13].

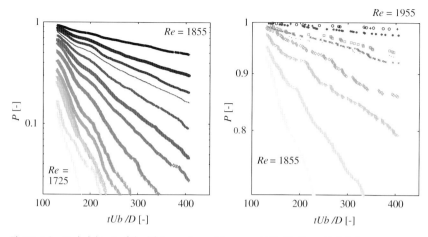

Figure 2.8 Probability P of the lifetime of a puff to exceed tU_b/D. From Ref. [14].

desired Reynolds number. The probability distribution of the lifetimes was determined based on 40–60 simulations per Reynolds number. They claimed that a quantitative agreement was found between their results and the previously obtained results by Peixinho and Mullin [12]. Although they concluded to also have found a linear scaling for the characteristic lifetime, the critical Reynolds number was 1870, which is significantly larger than the value of 1750 found by Peixinho and Mullin [12].

Meanwhile, another experiment was conceived by Hof et al. [15]. Rather than following disturbances in a pipe at fixed Re to determine the characteristic lifetime τ, one can consider a pipe with fixed length L and determine the probability that a disturbance survives: $P(t, Re; L)$. This is equivalent to considering the probability along vertical lines in Figure 2.5a (as opposed to evaluating $P(t, Re)$ along horizontal lines). A disturbance is introduced in a pipe (after sufficient distance from the pipe inlet) and it is observed when this disturbance reaches the outlet of the pipe. When the jet emanating from the pipe dips (as a result of the lower centerline velocity in the puff), the disturbance has obviously survived for the time period required to travel a distance L along the pipe. A schematic of this experiment is shown in Figure 2.9.

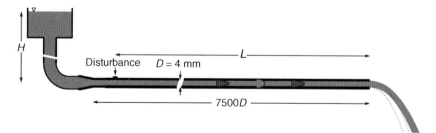

Figure 2.9 Schematic of the experiment used by Hof et al. [15].

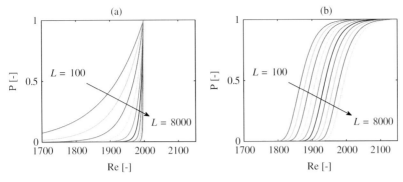

Figure 2.10 (a) Probability as a function of Reynolds number if $\tau^{-1} \sim (Re_c - Re)$, with $Re_c = 2000$. (b) Probability as function of Reynolds number if $\tau^{-1} \sim \exp(Re)$.

It was found from the analysis of the initial results reported by Faisst and Eckhardt [11] that the slope of the exponential distributions did not exactly correspond to the reported lifetimes that were based on the median lifetime extracted from Figure 2.5. Further analysis conjectured that the lifetime might actually scale as $\tau^{-1} \sim \exp(Re)$ [13]. This has an important consequence that the lifetime does not diverge at a finite critical Reynolds number, and underlines an important fundamental issue in regard to the transition to turbulence.

The difference between these two scaling regimes can be readily observed in $P(t, Re; L)$, as indicated in Figure 2.10. In the case of a linear scaling of τ^{-1} with a divergence of the lifetimes at finite Re_c, the observed probabilities for fixed L have an exponential shape, culminating at $P = 1$ for $Re = Re_c$. While for the exponential scaling, the probability curves have distinct S-shapes that shift to higher Re for increasing pipe length.

Note that the curves for $P(t, Re; L)$ for low probabilities ($P < 0.3$) look very similar. This implies that it is difficult to determine the difference between the two scaling regimes when data are available only for low Reynolds numbers and short pipes or observation times. This is a serious complication for numerical investigations, as not only the required integration time but also the computational cost increases with Reynolds numbers.

The linear scaling leads to an interesting thought experiment as the probability approaches a step function when $L \to \infty$. Consider a very long pipe driven by a constant pressure with the Reynolds number of the flow just below Re_c. Each disturbance introduced at the beginning of the pipe decays and for all disturbances the flow at the pipe outlet remains undisturbed. (The very long pipe length implies that the flow rate as determined by the pressure drop is not affected by introducing the disturbance.) Then a second, identical pipe is placed next to the first one. However, it is made slightly shorter, so that the flow rate increases just above the critical Reynolds number. When the same disturbance is introduced into both pipes, all disturbances in the second pipe will survive. Provided that the pipe are long enough, the difference in the two pipe lengths can be made arbitrarily small, so that the two pipes appear to be identical, yet their behavior in terms of the disturbances at the pipe exits is completely different.

In the case of the exponential scaling of the lifetime, the two (almost identical) pipes behave almost identically, with the slightly longer pipe having a slightly lower fraction of puffs surviving all the way to the pipe exit.

In order to make a distinction between the two scaling regimes, it is necessary to perform the measurements in very long pipes, preferably exceeding 1000–2000 pipe diameters in length. First a 4 mm diameter pipe with a length of 11 m ($L/D = 2,750$) was constructed in Delft and later a 4 mm diameter pipe with a total length of 30 m (i.e., $L/D = 7500$) was constructed in Manchester. The results of the measurements by Hof et al. [15] in both pipes are shown in Figure 2.11. Note that the probability curves have a distinct S-shape, which already indicates qualitatively that the lifetime of the disturbances does not diverge at finite Re. Further analysis indicated that the data indeed show an exponential scaling of the lifetime as a function of Re over the range of Reynolds numbers investigated.

Further experiments were conducted to extend the range of lifetimes that could be measured and to determine the location of decay quantitatively. Rather than considering the median lifetime, the rate of decay was determined. Thus, it was possible to determine τ for pipe lengths that were shorter than $U\tau$ (where U is the mean flow velocity) and it was possible to determine the decay rate over a very large range between 1 and 10^8. The results of the very large range in escape rates were reported by Hof et al. [16]. Later on, these results were confirmed by a quantitative measurement of the lifetime as reported by Kuik et al. [14]. The results for the lifetime, namely, decay rate or escape rate, as a function of Re are reproduced in Figure 2.12, together with the

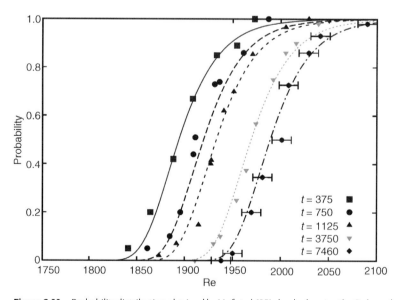

Figure 2.11 Probability distribution obtained by Hof et al. [15] clearly showing the S-shaped curves.

previous experimental and numerical results. It appears that the scaling is not exponential, but rather superexponential, that is, the lifetime τ is given by [16]

$$\tau^{-1} = (U/D)\exp[-(Re/c)^n], \quad (2.1)$$

with $c = 1549$ and $n = 9.95$, where n is related to the rate at which the laminar basin of attraction shrinks with increasing Reynolds number [17]. The data fit quite well with the experimental and numerical data obtained by others, except for very low Reynolds numbers and for Reynolds numbers where the observation is limited by the actual pipe length. The superexponential scaling also appears in the so-called spatiotemporal chaotic systems, where transients increase superexponentially with the size of the system [17].

The conclusion of the experiments is that the lifetimes of localized disturbances rapidly increases with Re. The scaling of the lifetime with Re as measured in the experiments suggests that the divergence of the lifetime does not occur at a finite critical Reynolds number. This suggests that, for the Reynolds numbers investigated, no evidence is found for the existence of a strange attractor. If the scaling found in the experiment holds for larger Re than those investigated, then it would imply that the turbulent flow state should be considered as a transient, albeit an extremely long-living one [18]. For example, to extend the observation time for the data shown in Figure 2.12 that are valid for water flowing through a $D = 1$ cm pipe to a Reynolds number of 2200 would imply observation times exceeding the lifetime of the universe. Evidently, extrapolation of the experimental results to any value of the Reynolds number beyond those investigated experimentally should be done with great caution. The problem is that the predicted lifetimes for higher Reynolds number become simply impractically large. In addition, at higher Reynolds numbers the localized nature of the turbulence is lost. At Reynolds number higher than approximately 2350, turbulent puffs can split and at Re higher than 2700 they can

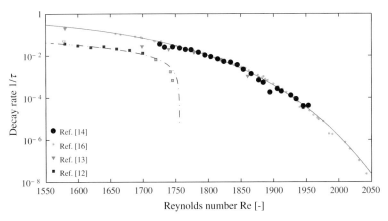

Figure 2.12 The lifetime of decay rate for $P(t)$ as a function of Reynolds number. Adapted from Ref. [14].

form "slugs" [2, 19]. This behavior cannot be explained from the present dynamical systems point of view on the transition to turbulence. Another remaining question is what the relation is between the exact periodic solutions that were found numerically (TWs) and the turbulent puff.

References

1 Reynolds, O. (1883) An experimental investigation of the circumstances which determine whether the motion of water shall be direct or sinuous, and of the law of resistance in parallel channels. *Philos. Trans. R. Soc.*, **174**, 935.

2 Wygnanski, I.J. and Champagne, F.H. (1973) On transition in a pipe. Part 1. The origin of puffs and slugs and the flow in a turbulent slug. *J. Fluid Mech.*, **59**, 281–335.

3 Drazin, P.G. and Reid, W.H. (2004) *Hydrodynamic Stability*, Cambridge University Press.

4 Faisst, H. and Eckhardt, B. (2003) Traveling waves in pipe flow. *Phys. Rev. Lett.*, **91** (22), 224502.

5 Wedin, H. and Kerswell, R.R. (2004) Exact coherent structures in pipe flow: travelling wave solutions. *J. Fluid Mech.*, **508**, 333–371.

6 Hof, B., van Doorne, C.W.H., Westerweel, J., Nieuwstadt, F.T.M., Faisst, H., Eckhardt, B., Wedin, H., Kerswell, R.R., and Waleffe, F. (2004) Experimental observation of nonlinear traveling waves in turbulent pipe flow. *Science*, **305** (5690), 1594–1598.

7 Pringle, C.T.P. and Kerswell, R.R. (2007) Asymmetric, helical, and mirror-symmetric traveling waves in pipe flow. *Phys. Rev. Lett.*, **99** (7), 074502.

8 Skufca, J.D., Yorke, J.A., and Eckhardt, B. (2006) Edge of chaos in a parallel shear flow. *Phys. Rev. Lett.*, **96** (17), 174101.

9 Schneider, T.M., Eckhardt, B., and Yorke, J.A. (2007) Turbulence transition and the edge of chaos in pipe flow. *Phys. Rev. Lett.*, **99** (3), 0034502.

10 Robert, C., Alligood, K.T., Ott, E., and Yorke, J.A. (2000) Explosions of chaotic sets. *Physica D*, **144** (1–2), 44–61.

11 Faisst, H. and Eckhardt, B. (2004) Sensitive dependence on initial conditions in transition to turbulence in pipe flow. *J. Fluid Mech.*, **504**, 343–352.

12 Peixinho, J. and Mullin, T. (2006) Decay of turbulence in pipe flow. *Phys. Rev. Lett.*, **96** (9), 094501.

13 Willis, A.P. and Kerswell, R.R. (2007) Critical behavior in the relaminarization of localized turbulence in pipe flow. *Phys. Rev. Lett.*, **98** (1), 014501.

14 Kuik, D.J., Poelma, C., and Westerweel, J. (2010) Quantitative measurement of the lifetime of localized turbulence in pipe flow. *J. Fluid Mech.*, **645** (1), 529–539.

15 Hof, B., Westerweel, J., Schneider, T.M., and Eckhardt, B. (2006) Finite lifetime of turbulence in shear flows. *Nature*, **443** (7107), 59–62.

16 Hof, B., Lozar, A., Kuik, D.J., and Westerweel, J. (2008) Repeller or attractor? Selecting the dynamical model for the onset of turbulence in pipe flow. *Phys. Rev. Lett.*, **101** (21), 214501.

17 Tél, T. and Lai, Y.-C. (2008) Chaotic transients in spatially extended systems. *Phys. Rep.*, **460**, 245–275.

18 Lathrop, D.P. (2006) Fluid dynamics: turbulence lost in transience. *Nature*, **443** (7107), 36–37.

19 Nishi, M., Bülent, Ü., Durst, F., and Gautam, B. (2008) Laminar-to-turbulent transition of pipe flow through puffs and slugs. *J. Fluid Mech.*, **614**, 425–446.

3
Dynamics of Patterns in Lasers with Delayed Feedback
Kirk Green and Bernd Krauskopf

3.1
Introduction

Modern semiconductor lasers are small, easy to produce in large numbers, and very efficient in terms of converting electrical energy into coherent light. These attractive properties are the reason that semiconductor lasers are today used in numerous technological applications, including optical communication networks and optical storage systems. On the negative side, semiconductor lasers are very susceptible to external optical influences, especially in the form of delayed feedback due to reflections from external optical elements and/or (bidirectional) coupling to other lasers. The delay arises due to the travel time of the light before it (re)enters the laser. Compared to the very fast timescales of the dynamics inside a semiconductor laser (on the order of picoseconds), distances between optical components on the order of centimeters already result in considerable delays that cannot be neglected; see, for example, Refs [1, 2] as entry points to the extensive literature.

We are concerned here with the simplest case of a semiconductor laser with conventional optical feedback (COF) in the form of reflections from a standard mirror at some distance from the laser. In fact, even very small amounts of COF, on the order of 0.1% of the output light, may destabilize the laser [3, 4]. This is why in practical applications expensive optical isolators must be employed when a semiconductor laser is coupled to optical fibers or other elements. The current classical case is that of a single-mode edge-emitting laser (EEL) subject to COF. In an EEL – the most commonly used type of semiconductor lasers today – light is produced in an active region in the shape of a one-dimensional waveguide (with a length of several hundred micrometers up to millimeters); the light exits at one (or both) of the laser's side facets, which act as (semitransparent) mirrors (see Figure 3.1a). Despite being quite long in the direction of lasing, single longitudinal mode operation of EELs can be ensured, for example, by incorporating internal frequency selective elements into the EEL. What is more, most EELs have a quite narrow active region (of a few micrometers) so that they generally lase at a single transverse mode. This means that the pattern of laser light at the exit facet consists of a single spot under all operating

The Complexity of Dynamical Systems. Edited by J. Dubbeldam, K. Green, and D. Lenstra
Copyright © 2011 WILEY-VCH Verlag GmbH & Co. KGaA, Weinheim
ISBN: 978-3-527-40931-0

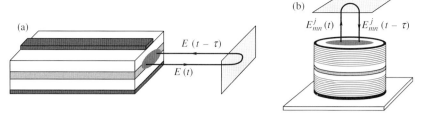

Figure 3.1 Sketches of an EEL with COF (a) and a VCSEL with COF (b).

conditions. Because of these properties, an EEL can be described in a rate equation approach by differential equations for the evolution of the (complex-valued) electric field and the density of charge carriers (or inversion) inside the laser. In the presence of COF, the electric field is coupled back to itself after a single fixed delay τ, and this results in the now famous Lang–Kobayashi (LK) equations [5]. This delay differential equation (DDE) has been shown in numerous studies to describe experimental measurements very well (for weak to moderate COF); see, for example, Refs [3, 6]. In general, there is an extensive literature on laser systems that feature delays due to (different types of) reflections from other optical elements and coupling between spatially separated lasers; see, for example, Refs [1, 2] and Chapter 4 and references therein.

The question arises what dynamics may ensue in the presence of COF when the laser supports more than one pattern of light at its exit facet. An example of such a laser is a broad-area EEL, where the active region is so wide that the light generally exits the laser not as a single wide spot, but as a one-dimensional pattern of light that consists of a number of bright spots; see, for example, Refs [7, 8]. We do not discuss broad-area lasers here, but rather focus on the case that the aperture of the laser is truly two dimensional. That is, we consider vertical-cavity surface-emitting lasers (VCSELs) – an increasingly important type of semiconductor laser with a cylindrical geometry, where a very thin, spatially extended active region is located between two stacks of mirrors. A VCSEL is pumped electrically via a circular contact and its light exits at the top face from a disk-shaped aperture (see Figure 3.1b). VCSELs are even more efficient than EELs and they are more easily integrated (on-chip) into larger arrays or optical circuits. Another advantage is that VCSELs provide consistent single longitudinal mode operation without additional elements (due to the very thin active region). On the other hand, because of the spatial extent of their active region and corresponding aperture, a VCSEL may support several transverse optical modes. In other words, the light does not always exit the VCSEL in a single large spot but generally in a two-dimensional pattern of bright spots on the disk-shaped aperture. The number of transverse optical modes, or patterns of light, that a VCSEL supports increases with the diameter of the aperture [7, 9]. In effect, one is dealing with a pattern formation problem, where the boundary conditions are given by the wave-guide properties of the VCSEL cavity. When the pump current is increased in a given VCSEL, the trivial off-solution becomes unstable and a first pattern of light emerges at the laser threshold. When the pump current is increased further, one typically finds a sequence of transitions to other and more complicated patterns. The patterns of light at the facet of the VCSEL are generally very close to the pure spatial modes of an equivalent passive waveguide. Due to the underlying circular symmetry of the VCSEL aperture, one finds patterns with the (approximate) discrete symmetries of a regular n-gon, where the number n increases with increasing pump current. Such patterns are generic and can also be found in other systems, such as in circular flame patterns [10].

We consider here a VCSEL with COF in the presence of more than one transverse optical mode. As opposed to the case of edge-emitting lasers, the modeling of VCSELs is still in development and under some dispute. However, due to the two-dimensional nature of the active region, a VCSEL needs to be described mathematically by partial differential equations for the electric field and the inversion, and we follow the modeling approach that has been used in Ref. [11]. When subject to COF, the overall model takes the form of a delayed partial differential equation (DPDE). Note that DPDE models have recently also been considered in other application areas, notably when spatial patterns or modes need to be controlled; examples include the control of Faraday wave patterns [12, 13] and dynamic testing methods for spatially extended mechanical systems such as beams [14]. To analyze or even simulate a DPDE, one generally needs to resolve the spatial part of the system with a suitable expansion method (e.g., by spatial discretization or Galerkin projection). After truncation, this leads to a (possibly quite large) system of DDEs. In Ref. [15], we presented an eigenfunction expansion method (EEM) that exploits the physical properties of the VCSEL in order to obtain a DDE that describes the spatiotemporal dynamics of the transverse modes under the influence of COF.

For the case of a VCSEL with COF that supports only two rotationally symmetric transverse optical modes, the EEM-reduced DDE is of quite low dimension, yet still describes the dynamics of the system accurately. This makes it possible to perform a bifurcation analysis with advanced numerical continuation tools of constant intensity solutions, called external cavity modes (ECMs), and even of oscillating intensity solutions. As a specific example of this new capability, we study here how the amount of self-feedback versus cross-feedback of the two spatial modes influences stability regions of ECMs and periodic solutions in the plane of feedback phase and feedback strength of the light as it reenters the laser. This is of practical interest because it is very difficult experimentally to determine how much the spatial modes influence each other via a COF loop. Our starting point is the observation that, in the degenerate case of pure self-feedback, the ECM structure of the two-mode VCSEL with COF is effectively that of a single-mode laser with COF. We then consider the transition to pure cross-feedback between the two modes. More specifically, we present how the two-parameter bifurcation diagram (in the plane of feedback phase and feedback strength) changes with the amount of self-feedback versus cross-feedback. Each bifurcation diagram consists of codimension-one bifurcation curves that meet and interact at points of higher codimension. Qualitative changes in the bifurcation diagrams are identified in the transition from pure self-feedback to pure cross-feedback. Overall, we find a number of intermediate regimes that differ so much that they might be used for characterizing the nature of the feedback in an experiment.

The chapter is organized as follows. Section 3.2 summarizes the single-mode COF laser as described by the LK equations. The EEM-reduced DDE model of a two-mode VCSEL is introduced in Section 3.3. The bifurcation analysis of its ECMs is discussed in Section 3.4, and the stability of bifurcating periodic solutions is discussed in Section 3.5. Section 3.6 summarizes and discusses directions for future research.

3.2
Single-Mode Laser with COF

In numerous studies, it has been shown that the single-mode COF laser is described well by rate equations for the complex electric field E and the carrier population N; see, for example, Refs [3, 6]. These rate equations have been introduced by Lang and Kobayashi [5], and they can be written in dimensionless form as

$$\frac{dE}{dt} = (1 + i\alpha)EN + \kappa e^{iC_p} E(t-\tau), \tag{3.1}$$

$$T\frac{dN}{dt} = P - N - (1 + 2N)|E|^2, \tag{3.2}$$

Here P is the pump current, T is the ratio between carrier and photon decay times, and α is the linewidth enhancement factor. The last term in Equation 3.1 models the optical feedback from the external mirror; here, the delay τ is given as the fixed external round-trip time, κ is the (real-valued) feedback strength, and C_p is the feedback phase (describing the subwavelength interaction between $E(t)$ and $E(t-\tau)$). The feedback phase C_p is used as one of our bifurcation parameters. In the representation of the results, it is helpful that C_p is 2π-periodic, which can be expressed as its invariance under the translation

$$C_p \to C_p + 2\pi. \tag{3.3}$$

Note that C_p has been shown to be accessible experimentally, allowing for an excellent comparison with numerical solutions of the LK equations (in the short external cavity regime) [6].

Because of the feedback term, system (3.1)–(3.2) is a DDE with a single fixed delay and, as such, it has as its phase space the infinite-dimensional space $C[-\tau, 0]$ of continuous functions with values in (E, N)-space. This is due to the fact that an entire history over the interval $[-\tau, 0]$ needs to be known to determine the future evolution of Equations 3.1 and 3.2. In particular, this means that already the basic stability analysis of solutions of DDEs is much more involved than that for (the nondelayed case of) ordinary differential equations (ODEs). That is, equilibria and periodic solutions of Equations 3.1 and 3.2 have stability spectra consisting of infinitely many eigenvalues [16]. Fortunately, it is possible to perform a numerical bifurcation analysis of a given DDE with the recently developed software tools DDE-BIFTOOL [17] and PDDE-CONT [18]; see also the survey papers [19, 20].

The basic solutions of Equations 3.1 and 3.2 are known as CW states or ECMs, and they are of the form

$$(E, N) = (R_s e^{i\omega_s t}, N_s), \quad (3.4)$$

where $R_s, \omega_s, N_s \in \mathbb{R}$. The ECMs arise due to the underlying S^1-symmetry of Equations 3.1 and 3.2 that is given by multiplication of the electric field with a complex number of modulus 1, that is, by the transformation $E \mapsto cE$ for any $c \in \mathbb{C}$ with $|c| = 1$; mathematically, an ECM is a group orbit under this symmetry group [21]. (Since R_s, ω_s, and N_s are fixed, ECMs are also referred to as fixed points in some parts of the literature.) To study the ECMs of the COF laser, one substitutes the ansatz (3.4) into the governing equations 3.1 and 3.2. Real and imaginary parts are then separated, yielding the ECMs as solutions of the transcendental equation

$$\omega_s = \kappa\sqrt{1 + \alpha^2} \sin\left(C_p - \omega_s \tau + \arctan(-\alpha)\right); \quad (3.5)$$

see also Refs [22–24]. Given the ECM frequency ω_s, the corresponding values for the carrier population N_s and amplitude R_s can be computed from

$$N_s = -\kappa \cos(C_p - \omega_s \tau + 2n\pi), \quad R_s^2 = \frac{P - N_s}{1 + 2N_s}. \quad (3.6)$$

Analytical results about the existence of ECMs (and some results about their stability) can be derived from Equations 3.5 and 3.6; for example, turning points correspond to saddle-node bifurcations of ECMs; see Refs [22, 24] for details and Ref. [25] as an entry point to asymptotic methods.

In order to continue ECMs in parameters and to determine their stability, it is necessary to resolve their S^1-symmetry so that every ECM is an isolated solution (rather than a circle in (E, N)-space). This can be achieved by moving to a rotating frame of reference with fixed frequency b, as given by

$$E \to E e^{ibt}, \quad b \in \mathbb{R}. \quad (3.7)$$

After inserting Equation 3.7 into Equation 3.1, an ECM can be studied as a steady-state solution for which $b \equiv \omega_s$; see Refs [19, 26] for further details. Figure 3.2 shows the two-parameter bifurcation diagram of ECMs in the (C_p, κ)-plane, where we fix the laser parameters at $P = 8.0$, $T = 750$, and $\alpha = 3.0$; furthermore, we consider a fixed delay time of $\tau = 500$, which corresponds to a distance of approximately 10 cm between the laser and the mirror. Figure 3.2a shows the bifurcation diagram in the covering space over four periods of C_p, so that the 2π-translational symmetry (3.3) is clearly visible. The image in Figure 3.2b shows the bifurcation diagram for $C_p \in [-\pi, \pi]$, that is, over the fundamental domain of (3.3). The bifurcation diagram has been computed with the package DDE-BIFTOOL, and it shows curves S of saddle-node bifurcations and curves H of Hopf bifurcations. These curves divide the (C_p, κ)-plane into regions with different numbers of ECMs. In the light gray region for small κ there is one ECM, which is stable and the continuation of the solitary laser solution (that exists for $\kappa = 0$). Once κ is chosen above the cusp point C, new ECMs

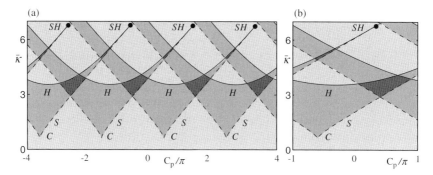

Figure 3.2 Two-parameter bifurcation diagram of ECMs of the LK equations 3.1 and 3.2 in the (C_p, κ)-plane. While (a) shows C_p over a wide range to bring out the 2π translational symmetry, (b) shows it over one period in the interval $[-\pi, \pi]$. Shown are curves S of saddle-node bifurcation, curves H of Hopf bifurcation, and a codimension-two cusp point C, and a codimension-two saddle-node Hopf point SH; black curves correspond to supercritical and gray curves correspond to subcritical bifurcations. Gray shading of increasing intensity indicates the number of simultaneously stable ECMs, here up to 3; $P = 8.0$, $T = 750$, $\alpha = 3.0$, $\tau = 500$, and $\bar{\kappa} = \kappa \times 10^3$.

may be born in pairs in saddle-node bifurcations, one of which may be initially stable. Stable ECMs lose their stability for increasing κ in Hopf bifurcations, which give rise to stable oscillations of the laser power at the characteristic relaxation oscillation frequency, on the order of gigahertz. Note that the curves S and H meet at the point SH, which is known as a codimension-two saddle-node Hopf bifurcation [27, 28]. At this point, there is a change in the character of the curves S and H from super- to subcritical, meaning that they correspond to bifurcations from which emanate unstable solutions for κ-values above SH. For a detailed bifurcation analysis of the ECM structure of the COF laser, see Refs [22, 24].

3.3
VCSEL with Optical Feedback

A VCSEL may support different patterns of light on its aperture. Under the assumption of weak guiding, these patterns can be described mathematically as linearly polarized (LP) optical modes, which are roots of the Helmholtz equation [11, 29]. The resulting modes LP_{mn} can be expressed in terms of Bessel functions; they are characterized by n maxima of the Bessel function in the radial direction, and $2m$ zeros in the azimuthal direction of the cylindrical waveguide; some examples of LP modes are shown in Figure 3.3.

Following Refs [11, 29–31], a VCSEL can be described in a rate equation approach by PDEs for the evolution of the slowly varying complex electric fields $E_{mn}^j(t)$ of the modes LP_{mn} with polarization j, coupled via the real-valued spatial carrier population $N(r, t, \phi)$. (Here, the polarization index j can be either c or s, which denote the two orthogonal cosine and sine instances of the mode LP_{mn}; note that the modes LP_{0n} are

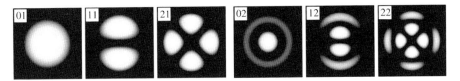

Figure 3.3 Linearly polarized optical modes LP$_{mn}$ of a VCSEL.

rotationally symmetrical, so that they are not distinguished in terms of transverse polarization.) In dimensionless form, one obtains

$$\frac{dE_{mn}^j(t)}{dt} = (1+i\alpha)\xi_{mn}^j(t)E_{mn}^j(t) + F_{mn}^j(t), \tag{3.8}$$

$$T\frac{\partial N(r,\phi,t)}{\partial t} = \frac{d_f}{r}\frac{\partial}{\partial r}\left[r\frac{\partial N(r,\phi,t)}{\partial r}\right] + \frac{d_f}{r^2}\frac{\partial^2 N(r,\phi,t)}{\partial \phi^2} + P(r,\phi,t)$$
$$- N(r,\phi,t) - \sum_n\left((1+2\xi_{0n}(t))|E_{0n}(t)|^2|\Psi_{0n}(r,\phi)|^2\right. \tag{3.9}$$
$$+ \sum_m\sum_{j=c,s} 2(1+2\xi_{mn}^j(t))|E_{mn}^j(t)|^2|\Psi_{mn}^j(r,\phi)|^2\right).$$

The spatial pump $P(r,\phi,t) \equiv P(r)$ describes carrier injection over the core region of the VCSEL via a rotationally symmetric contact. The modal gains ξ_{mn}^j are given by integrals describing the spatial overlap between the respective optical field and the inversion; see Ref. [15] for details. The functions $F_{mn}^j(t)$ describe an external perturbation that is applied to the electric fields of the modes LP$_{mn}$. In the present case, this external perturbation is optical feedback from a mirror at a fixed distance, meaning that Equations 3.8 and 3.9 take the form of a delayed PDE. To make this system amenable to a full nonlinear bifurcation analysis, one needs to resolve the spatial dependence. As is common [11], the azimuthal direction can be resolved through the use of a Fourier series expansion

$$N(r,\phi,t) = \sum_{k=0}^{\infty}(N_{ck}(r,t)\cos(k\phi) + N_{sk}(r,t)\sin(k\phi)). \tag{3.10}$$

The radial direction could be resolved via the use of a finite difference discretization scheme [11, 31]. However, this typically results in a large-scale system of differential equations that are difficult to handle both analytically and numerically. Therefore, we choose to perform a second eigenfunction expansion. That is, we use a kth-order Bessel function expansion [15]

$$N_k(r,t) = \sum_{q=1}^{\infty} N_{kq}(t)J_k(\gamma_{k,q}r). \tag{3.11}$$

After these two expansions, and after suitable orthogonality conditions have been applied [15], one is left with a spatially resolved system describing a multitransversemode VCSEL (with optical feedback). Importantly, the integral functions that describe the spatial overlaps between the inversion density with the pump current and the electric field, respectively, take the form of constant coupling constants. Hence, they can be evaluated a priori, which leads to a considerable speedup in numerical computations; see Ref. [15] for details of the general case.

3.3.1
EEM-Reduced Model of a Two-Mode VCSEL with Optical Feedback

Here, we consider in detail the case that the VCSEL supports only the first two rotationally symmetric modes LP_{01} and LP_{02} (see Figure 3.3). These symmetric modes are desirable in optical communication schemes for their coupling efficiency, and they may be found stably in certain types of VCSELs, notably those with smaller aperture diameter [9]. The two-mode VCSEL has been studied experimentally in Ref. [32], and theoretical studies can be found in Refs [15, 31, 33]. From the mathematical point of view, the LP_{01} and LP_{02} modes can be described by their radial profile alone, which allows for a substantial reduction of the dimensionality of the overall rate equations. That is, one obtains the dimensionless reduced DDE

$$\frac{dE_1(t)}{dt} = (1+i\alpha)\xi_1 E_1(t) \\ + \kappa e^{iC_p}[\eta E_1(t-\tau) + (1-\eta)E_2(t-\tau)], \quad (3.12)$$

$$\frac{dE_2(t)}{dt} = (1+i\alpha)\xi_2 E_2(t) \\ + \kappa e^{iC_p}[\eta E_2(t-\tau) + (1-\eta)E_1(t-\tau)], \quad (3.13)$$

$$T\frac{dN_q(t)}{dt} = -(\gamma_{0,q}^2 d_f + 1)N_q(t) + \varrho_q \\ - \sum_{n=1}^{2}((1+2\xi_n)|E_n(t)|^2 \beta_n^{0q}). \quad (3.14)$$

We consider here a total of $q = 14$ expansion terms for the spatial variable N, which was found in previous studies [15, 31] to give sufficient accuracy. (For simplicity, we write $N_{kq}(t)$ in Equation 3.11 as $N_q(t)$.) Hence, the overall system consists of 18 differential equations in total. Here, $\gamma_{0,q}$ are the roots of the qth Bessel function J_k, and the overlap integrals (which need to be evaluated only once) are given by

$$\xi_n = \sum_{q=1}^{14} \int_0^1 (|\Psi_n(r)|^2 N_q J_0(\gamma_{0,q}r))r dr, \quad (3.15)$$

$$\varrho_q = \frac{2}{[J_1(\gamma_{0,q})]^2} \int_0^1 P(r) J_0(\gamma_{0,q}r)\, r dr, \quad (3.16)$$

$$\beta_n^{0q} = \frac{2}{[J_1(\gamma_{0,q})]^2} \int_0^1 |\Psi_n(r)|^2 J_0(\gamma_{0,q} r) \, r \, dr. \tag{3.17}$$

Furthermore, the spatial pump $P(r)$ in Equation 3.16 is modeled as

$$P(r) = \frac{P_{\max}(1 + \mathrm{erf}(2\sqrt{75}(-r + 0.3)))}{2}. \tag{3.18}$$

The optical feedback has been introduced explicitly into Equations 3.12 and 3.13 for the electric fields of the modes LP$_{01}$ and LP$_{02}$, respectively. As was the case for the LK equations in Section 3.2, for each of the two fields, the feedback has a common delay time τ, a common strength κ, and a common (2π-periodic) phase C_p. The feedback terms depend on the homotopy parameter $\eta \in [0, 1]$, which models the amount of self-feedback versus cross-feedback of the two modes. Specifically, for $\eta = 1$, both modes receive only their own feedback, which we refer to as pure self-feedback; in fact, before the study in Ref. [31], only pure self-feedback was assumed when modeling VCSELs with feedback. However, one would expect that the modes are no longer orthogonal in the far field (at the point of reflection), and/or defects in the external mirror's surface may result in coupling between the modes. In short, one would expect a certain amount of cross-feedback, that is, $\eta < 1$; note that the amount of self-feedback versus cross-feedback as modeled by η might be changed experimentally by mirror shaping or the use of frequency selective feedback [34]. The other extreme is that of pure cross-feedback for $\eta = 0$, where LP$_{01}$ receives feedback only from LP$_{02}$, and vice versa. Hence, by decreasing η from 1 to 0 we are able to study over the entire range the influence of self-coupling versus cross-coupling on the dynamics of Equations 3.12 and 3.13 as represented by the bifurcation diagram of ECMs in the (C_p, κ)-plane.

3.3.2
Analytical Results for the Two-Mode ECM Structure

The basic solutions of Equations 3.12–3.14 are ECMs, which are of the form

$$(E_1(t), E_2(t), N_{1,\ldots,14}(t)) = (R_1 e^{i\omega_s t}, R_2 e^{i\omega_s t + i\Phi}, N_{1,\ldots,14}), \tag{3.19}$$

where $R_1, R_2, \omega_s, \Phi, N_1, \ldots, N_{14} \in \mathbb{R}$. In other words, both modes have constant but generally different amplitudes R_1 and R_2, which both feed from the same constant reservoir of spatially distributed carriers as expressed by the constants $N_1, \ldots, N_{14}$ and Equation 3.11. Furthermore, both fields have the same frequency ω_s with a constant phase shift Φ between them. We remark that in reality there is a frequency difference between the two modes, but it is extremely small (only about 1 THz), which motivates modeling the frequencies of the two modes as identical.

The ECMs for the two-mode VCSEL as given by (3.19) again arise from the underlying S^1-symmetry of Equations 3.12–3.14. It is given by multiplication of *both* electric fields with any complex number of modulus 1, that is, by the transformations

$$(E_1, E_2) \mapsto (cE_1, cE_2) \quad \text{for } c \in \mathbb{C} \text{ with } |c| = 1. \tag{3.20}$$

As was the case for the LK equations, an ECM is a group orbit under this symmetry group; when projected onto the $(R_1, R_2, N_{1...14})$-space, an ECM can be treated as a steady-state solution [21].

As before, we substitute the ansatz (3.19) into Equations 3.12–3.14. This yields two coupled transcendental equations for the frequency ω_s and the phase difference Φ, which can be written as

$$\omega_s = \kappa\eta\sqrt{1+\alpha^2}\sin\left(C_p - \omega_s\tau + \arctan(-\alpha)\right)$$
$$+ \kappa(1-\eta)\frac{R_2}{R_1}\sqrt{1+\alpha^2}\sin\left(C_p - \omega_s\tau + \arctan(-\alpha) + \Phi\right), \quad (3.21)$$

$$\omega_s = \kappa\eta\sqrt{1+\alpha^2}\sin\left(C_p - \omega_s\tau + \arctan(-\alpha)\right)$$
$$+ \kappa(1-\eta)\frac{R_1}{R_2}\sqrt{1+\alpha^2}\sin\left(C_p - \omega_s\tau + \arctan(-\alpha) - \Phi\right). \quad (3.22)$$

Even though these formulas give a way of eliminating R_1 and R_2, unfortunately, it is not possible to derive a general formula for just ω_s or just Φ. (Alternatively, one could eliminate Φ but then R_1 and R_2 remain.) This situation is analogous to the task of finding the compound laser modes of two mutually delay-coupled lasers [35].

However, we can analyze the ECM structure for the two special cases of pure self-feedback where $\eta = 1$ and of pure cross-feedback where $\eta = 0$. First, for $\eta = 1$ the second terms of (3.21) and (3.22) are zero, so that both equations reduce exactly to Equation 3.5. Hence, the equation for the ECMs for the two-mode VCSEL with pure self-feedback is exactly the equation for the ECMs of the COF laser. Notice, however, that for the two-mode VCSEL model Equation 3.5 represents two identical solutions, which constitutes a degenerate (double-covered) situation. That is, for $\eta = 1$ the constant Φ shift between E_1 and E_2 can take any value, and one can speak of a Φ-indeterminacy.

For $\eta = 0$, on the other hand, the first terms of (3.21) and (3.22) are zero, so that the two transcendental equations reduce to

$$\omega_s = \kappa\frac{R_2}{R_1}\sqrt{1+\alpha^2}\sin\left(C_p - \omega_s\tau + \arctan(-\alpha) + \Phi\right), \quad (3.23)$$

$$\omega_s = \kappa\frac{R_1}{R_2}\sqrt{1+\alpha^2}\sin\left(C_p - \omega_s\tau + \arctan(-\alpha) - \Phi\right). \quad (3.24)$$

By eliminating R_1 and R_2, we obtain

$$(\omega_s)^2 = \kappa^2(1+\alpha^2)\sin\left(C_p - \omega_s\tau + \arctan(-\alpha) + \Phi\right)$$
$$\times \sin\left(C_p - \omega_s\tau + \arctan(-\alpha) - \Phi\right). \quad (3.25)$$

We cannot solve this coupled system, but we can conclude that for $\eta = 0$ the ECMs satisfy the additional π-symmetries given by the translations

$$C_p \mapsto C_p + \pi \quad \text{and} \quad \Phi \mapsto \Phi + \pi. \quad (3.26)$$

In particular, this implies that the bifurcation diagram in the (C_p, κ)-plane is invariant under translation of C_p by π (and not only under translation by 2π). We remark that Equation 3.25 is very similar to that determining the compound laser modes of two mutually delay-coupled lasers with zero frequency detuning [35]. This suggests that the two-mode VCSEL with pure cross-feedback could be interpreted as two spatially extended, mutually delay-coupled VCSELs – one lasing at the LP_{01} mode and the other lasing at the LP_{02} mode, and both having the same free-running frequency. Another requirement is that the two VCSELs would have to have identical carrier dynamics (as they compete for the same carrier reservoir). This latter requirement constitutes the main difference with the case of two mutually delay-coupled *independent* lasers as studied in Ref. [35].

Overall, we find that the two extreme cases of pure self-feedback and of pure cross-feedback have different special properties. The main question that we will address next is how the corresponding bifurcation diagrams of ECMs in the (C_p, κ)-plane transform into one another as the homotopy parameter $\eta \in [0, 1]$ is changed.

3.4
Numerical Bifurcation Analysis of the Two-Mode ECM Structure

We now perform a numerical bifurcation analysis of the ECMs of Equations 3.12–3.14 with the continuation package DDE-BIFTOOL [17]. To this end, we again first resolve their S^1-symmetry (3.20) by moving to a rotating frame of reference with frequency b, which is given by

$$(E_1, E_2) \rightarrow (E_1 e^{ibt}, E_2 e^{ibt}), \tag{3.27}$$

so that the ECMs can be studied as steady-state solutions for which $b \equiv \omega_s$. The ECMs can be found and continued in parameters. To this end, we fix the material parameters at $P_{\max} = 2.0$, $T = 750$, and $\alpha = 3.0$; furthermore, we set the diffusion constant to $d_f = 0.05$. As before, we consider a fixed delay time of $\tau = 500$, and use the feedback phase C_p and the feedback strength κ as free parameters.

Motivated by the analysis in Section 3.3.2, our starting points are the bifurcation diagrams in the (C_p, κ)-plane of Equations 3.12–3.14 for $\eta = 1.0$ and for $\eta = 0.0$, which are shown in Figure 3.4. For the case of pure self-feedback of $\eta = 1.0$ in Figure 3.4a, we find a curve DZ of double-zero eigenvalues and a curve H_2 of Hopf bifurcations. The two curves meet at a point, labeled DZH_2, where they change from super- to subcritical, which is indicated by a change of the curves from black to gray. Notice also the cusp point C on the curve DZ. The gray shading indicates the number of coexisting stable ECMs in the different regions, of which there are up to 3. As was discussed in Section 3.3.2, the bifurcation diagram in Figure 3.4a is indeed as that for an equivalent COF laser; compare with Figure 3.2b. The difference is that instead of a saddle-node bifurcation (with a single eigenvalue zero) we find a double-zero eigenvalue along the curve DZ, which stems from the fact that we are dealing with a degenerate (double-covered) case for $\eta = 1.0$.

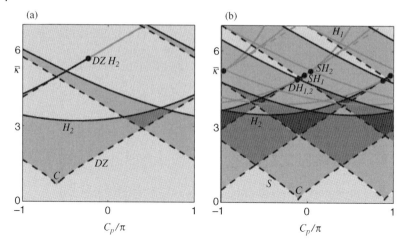

Figure 3.4 Two-parameter bifurcation diagrams of ECMs of Equations 3.12–3.14 in the (C_p, κ)-plane. (a) The double-covered case $\eta = 1.0$, with curves DZ of double-zero eigenvalues and a curve H_2 of Hopf bifurcations. (b) The case $\eta = 0.0$, with curves S of saddle-node bifurcation and curves H of Hopf bifurcation. Gray shading of increasing intensity indicates the number of simultaneously stable ECMs; $P_{max} = 2.0$, $T = 750$, $\alpha = 3.0$, $d_f = 0.05$, $\tau = 500$, and $\bar{\kappa} = \kappa \times 10^3$.

The bifurcation diagram for the case of pure cross-feedback for $\eta = 0.0$ is shown in Figure 3.4b. One immediately notices its additional symmetry of translation by π in C_p in accordance with (3.26). As a result, we find two curves S of saddle-node bifurcations, with two cusp point C in the C_p-interval $[-\pi, \pi]$. Similarly, we find two sets of Hopf bifurcation curves H_1 and H_2, which meet the curves S at codimension-two saddle-node Hopf bifurcations SH_1 and SH_2. In accordance with general results of bifurcation theory, the curves S, H_1, and H_2 change from super- to subcritical at these bifurcation points. The point marked $DH_{1,2}$ is a codimension-two double-Hopf bifurcation, where the curves H_1 and H_2 cross transversally. Gray shading again indicates the number of coexisting stable ECMs in the different regions, of which there are up to four.

3.4.1
Dependence of the Bifurcation Diagram on η

We now consider the transition between the two special cases in Figure 3.4 as the homotopy parameter η is changed. To give an initial impression, Figure 3.5 shows one-parameter bifurcation diagrams for fixed $\kappa = 0.003$ where the power of the first electric field $P_1 = R_1^2$ is plotted against the feedback phase C_p (over the fundamental interval $[-\pi, \pi]$). From (a) to (f), η is increased from $\eta = 0.0$ in intermediate steps to $\eta = 1.0$. Hence, panels (a) and (f) of Figure 3.5 correspond to one-dimensional cross sections at $\kappa = 0.003$ through the two panels of Figure 3.4, and the remaining panels illustrate the transition between these two cases. In Figure 3.5, parts of the branches are drawn black when the corresponding ECM is stable, and gray otherwise.

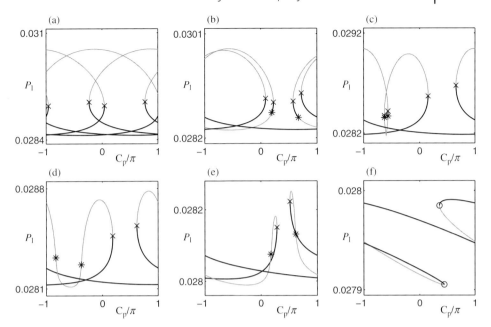

Figure 3.5 One-parameter bifurcation diagrams showing the power P_1 of LP_{01} against C_p. Stable parts of the branches are black and unstable part gray; marked are saddle-node bifurcations ($\times$), Hopf bifurcations ($*$), and in (f) degenerate double-zero points ($\circ$). From (a) to (f), η takes the values 0.0, 0.3, 0.6, 0.7, 0.9, and 1.0; $\kappa = 0.003$ and other parameters are as in Figure 3.4.

Furthermore, saddle-node bifurcations (fold points with respect to C_p) and Hopf bifurcations are marked.

For pure cross-feedback at $\eta = 0$, the ECM branches in Figure 3.5a exhibit the additional π-symmetry 3.26. Furthermore, for any value of C_p one finds either three or four stable coexisting ECMs, which agrees with the two-dimensional bifurcation diagram in Figure 3.4b when crossed at $\kappa = 0.003$. As η is increased from zero, the additional π-symmetry is immediately broken. Furthermore, one notices the emergence of Hopf bifurcations, which destabilize parts of the ECM branches. Already in Figure 3.5b for $\eta = 0.3$, the number of coexisting stable ECMs is reduced to at most two. Notice further that the branches have moved through one another, which leads to a C_p-interval with only one stable ECM. As η is increased further to $\eta = 0.6$ in Figure 3.5c, the Hopf bifurcations move closer to the saddle-node bifurcations near a small loop in the ECM branch. In Figure 3.5d for $\eta = 0.7$, one finds that this loop has disappeared together with the associated saddle-node bifurcations (i.e., fold points). For $\eta = 0.9$ in Figure 3.5e, the two Hopf bifurcation points move toward the remaining two saddle-node bifurcations. Finally, for pure self-feedback at $\eta = 1.0$ in Figure 3.5f, the Hopf bifurcations take place exactly at the fold points. In fact, these points are degenerate double-zero eigenvalue points; compare with Figure 3.4a.

Figure 3.5 raises some immediate questions. Where do the Hopf bifurcation points in Figure 3.5b come from? How do the two saddle-node bifurcation points disappear? Furthermore, what is the mechanism that produces the degenerate double-zero point in the limit $\eta = 1.0$? To answer these questions, we consider the changes of the entire two-parameter bifurcation diagram in the (C_p, κ)-plane with η. Figure 3.6 shows nine intermediate bifurcation diagrams in η-steps of 0.1, where the bifurcation curves and regions of ECM stability are represented as in Figure 3.4. In addition, we find a white region without stable ECMs, as well as a codimension-two

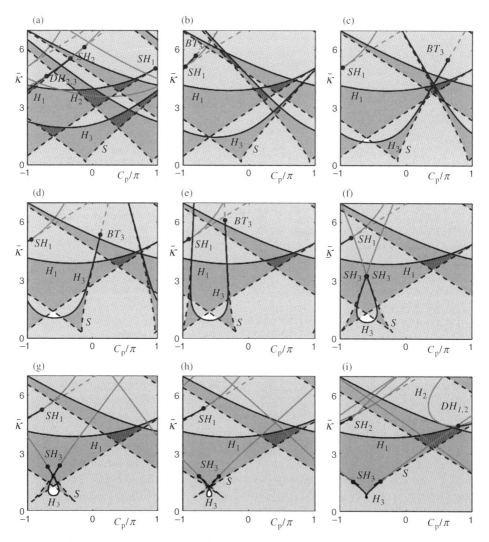

Figure 3.6 The two-parameter bifurcation diagrams of ECMs of Equations 3.12–3.14 in the (C_p, κ)-plane for increasing values of η. From (a) to (i), η takes the values 0.1, 0.2, 0.3, 0.4, 0.5, 0.6, 0.7, 0.8, and 0.9; other parameters are as in Figure 3.4.

Bogdanov–Takens bifurcation point BT_3 (corresponding to isolated double-zero eigenvalues [27, 28]).

Again, a first conclusion from Figure 3.6 is that, as η is increased from zero, the additional π-symmetry in C_p is immediately lost (see Figure 3.6a for $\eta = 0.1$). In particular, a difference emerges of how the (two sets of) saddle-node and Hopf curves interact. That is, already for $\eta = 0.3$ as in Figure 3.6c there is a single codimension-two saddle-node Hopf point SH_1 near $C_p = -\pi$, and a single Bogdanov–Takens point BT_3 near $C_p = 0.7\pi$; details of this transition near $\eta = 0.0$ are presented in Section 3.4.2. There is an important distinction between the corresponding Hopf bifurcation curves H_1 and H_3: the ensuing periodic solutions are of two different types. That is, when H_1 is crossed well-known relaxation oscillations (ROs) arise. Crossing H_3, on the other hand, results in the emergence of external-cavity oscillations (EOs) with a period of about the external round-trip time τ [15]. Notice further from Figure 3.6 that, as η is increased, the Hopf curve H_3 associated with the EOs moves toward lower values of κ. As a result, one would expect to find EOs already for lower values of κ compared to ROs; the stability regions and the bifurcations of ROs and EOs are discussed in detail in Section 3.5.

As the curve H_3 moves with η, a region without any stable ECMs opens up for $\eta \geq 0.2$. This occurs when H_3 dips below the crossing point of the saddle-node curves (see Figure 3.6b). Increasing η further, we find that the Hopf curve H_3 becomes steeper and changes slope (see Figure 3.6e). This is due to the fact that a point of self-intersection of H_3 moves into the chosen κ-range of $[0, 0.007]$, giving rise to a loop of H_3 (see Figure 3.6f). The Bogdanov–Takens point BT_3 has left our κ-region of interest, and instead a saddle-node Hopf point SH_3 has moved toward lower values of κ. Furthermore, a second saddle-node Hopf point SH_3 on H_3 also entered the κ-region of interest; in Figure 3.6f the two points SH_3 lie very close to one another, and they are responsible for the changes from super- to subcritical of both S and H_3. As η is increased further, the two sets of saddle-node bifurcation curves, and the associated cusps points, converge to one another. Simultaneously, the loop of the subcritical part of H_3 decreases in size and converges to the cusp point on S (see Figure 3.6g–i). Note that the bifurcation diagram in Figure 3.6i for $\eta = 0.9$ is a small perturbation of the degenerate double-covered case in Figure 3.4a for $\eta = 1.0$. Details of the transition process toward $\eta = 1.0$ are presented in Section 3.4.3.

3.4.2
Transitions Involving Codimension-two Points Near $\eta = 0.0$

We now consider in more detail what happens to the two sets of codimension-two points SH_1 and SH_2 in Figure 3.4b as η is increased. These two sets of points are identical for $\eta = 0.0$ (because of the additional π-periodicity of C_p), but they undergo two different transitions for $\eta > 0.0$. These are sketched in Figures 3.7 and 3.8. The starting point is the interaction for $\eta = 0.0$ of the saddle-node curve S and Hopf curves H_1, H_2, and H_3 in codimension-two double-Hopf, saddle-node Hopf and Bogdanov–Takens bifurcation points as sketched in Figure 3.7a. When compared with Figure 3.4b, we note that the point BT_3 in Figure 3.7a is outside the shown κ-range.

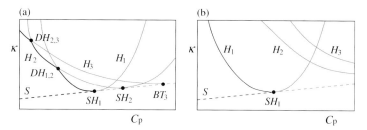

Figure 3.7 Sketch of the first type of interactions of double-Hopf DH, saddle-node Hopf SH, and Bogdanov–Takens BT bifurcation points as η is increased from zero.

Figure 3.7 shows bifurcations associated with the point SH_1 near $C_p = \pi$ in Figure 3.4b. The ensuing transition is quite simple: with increasing $\eta > 0$ the double-Hopf points $DH_{1,2}$ and $DH_{2,3}$ move rapidly up and out of our region of interest. At the same time, the points SH_2 and BT move up and to the right along the curve S. As a result, the Hopf curve H_1 is the only Hopf curve in the region of interest with a supercritical part, up to the point SH_1 (see Figure 3.7b). The sketched situation is similar to that near SH_1 in Figure 3.6.

The second transition, sketched in Figure 3.8, concerns bifurcations associated with the point SH_1 near $C_p = 0$ in Figure 3.4b. Starting from the same situation in Figure 3.8a, as η is increased from zero the two saddle-node Hopf points SH_1 and SH_2 move closer together and the double-Hopf point $DH_{1,2}$ moves down along H_1. The three points then all come together in a degenerate saddle-node Hopf bifurcation point, after which SH_1 and SH_2 exchange positions along S and $DH_{1,2}$ disappears (see Figure 3.8b). Next, the points SH_1 and BT_3 exchange their positions along S (see Figure 3.8c). This is the situation one finds for $\eta = 0.1$ in Figure 3.6a. When η is

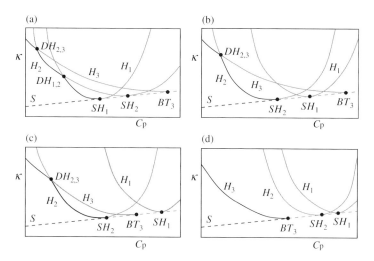

Figure 3.8 Sketch of the second type of interactions of double-Hopf DH, saddle-node Hopf SH, and Bogdanov–Takens BT bifurcation points as η is increased from zero.

increased above $\eta = 0.1$, the points SH_2 and BT_3 also exchange their positions along S, which involves the disappearance of the double-Hopf point $DH_{2,3}$ as well. As a result, the curve H_3 is the only curve with a supercritical part in the region of interest, and it ends at the Bogdanov–Takens point BT_3 (see Figure 3.8d). This is exactly the situation we find in Figure 3.6b–e, where the point BT_3 has entered the relevant κ-range while SH_1 and SH_2 lie outside this range.

The result of this, necessarily quite mathematical, discussion is a consistent picture near the special bifurcation diagram for $\eta = 0.0$, which involves several codimension-three bifurcations of ECMs in quick succession at η is increased.

3.4.3
Details of the Transition to $\eta = 1.0$

We now investigate the emergence of the loop of the Hopf bifurcation curve H_3 in Figure 3.6, its interaction with the saddle-node curve S, and the subsequent convergence to the degenerate bifurcation diagram for $\eta = 1.0$. The associated changes to the bifurcation diagram are shown in Figures 3.9 and 3.10. For $\eta \leq 0.4$, there are two distinct saddle-node curves. However, as the loop of H_3 develops, we find that these two curves connect (at infinity) and form a single curve S that has a cusp point at large values of κ (see Figure 3.9a). When η is increased further, one encounters a codimension-three swallowtail singularity [36] that creates two further cusp points on the right-hand side of this (subcritical) saddle-node curve (see Figure 3.9b). As η is

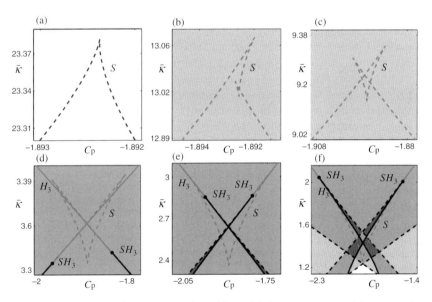

Figure 3.9 Emergence of cusp points on the saddle-node bifurcation curve S and their interaction with the Hopf curve H_3. From (a) to (f), η takes the values 0.51, 0.52, 0.53, 0.59, 0.63, and 0.75; other parameters are as in Figure 3.4.

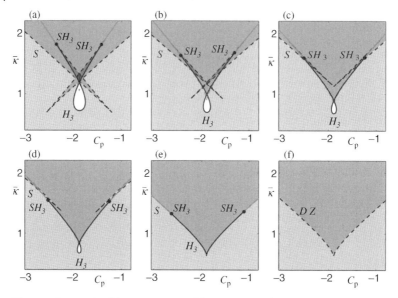

Figure 3.10 Details of the convergence of the supercritical loop of the Hopf curve H_3 to the degenerate case for $\eta = 1.0$. From (a) to (f), η takes the values 0.8, 0.85, 0.88, 0.9, 0.99, and 1.0; other parameters are as in Figure 3.4.

increased further, one of the cusp points moves through the left branch of S, forming what looks like a down-pointing arrow, while this entire structure moves to lower values of κ (see Figure 3.9c). For $\eta \approx 0.59$, the structure starts to interact with the Hopf curve H_3, meaning that the saddle-node Hopf bifurcations SH_3 enter into the region near the cusp points (see Figure 3.9d). Increasing η further, the points SH_3 move up along S, as in Figure 3.9e, and then move (one by one) past the upper cusp points onto the parts of S between the cusp points (see Figure 3.9f). As a result, most of the curve S (with the exception of the "arrow tip") is supercritical and, hence, part of the ECM stability boundary; compare with Figure 3.6g. Notice also that the entire structure is inside the region of interest for $\eta \geq 0.6$.

Figure 3.10 illustrates in detail how the structure consisting of the cusp points on S and the supercritical loop of H_3 develops further as the degenerate case $\eta = 1.0$ is approached. Figure 3.10a is an enlargement of the situation for $\eta = 0.8$. As η is increased, the lower cusp points move toward larger values of κ and above the lowest cusp point on S (see Figure 3.10b). The two sets of two remaining cusp points (on the right and the left), and the associated short parts of S between them that contain the saddle-node Hopf points SH_3, move closer to the each other (see Figure 3.10c and d). With increasing η, they disappear in swallowtail bifurcations, resulting in a situation as in Figure 3.10e for $\eta = 0.99$. Notice also that the entire curve H_3 is extremely close to S (the loop has shrunk down almost to the remaining cusp point on S). Indeed, the curves S and H for $\eta \approx 1.0$ are perturbations of the double eigenvalue zero curve DZ of the limit $\eta = 1.0$ in Figure 3.10f.

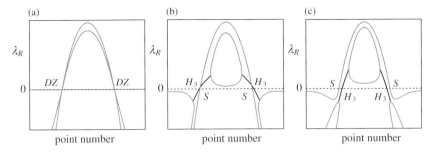

Figure 3.11 Continuation in C_p of the leading eigenvalues of an ECM, represented by their real parts as a function of the DDE-BIFTOOL point number; gray curves represent real eigenvalues and black curves complex-conjugate eigenvalues. (a) For $\eta = 1.0$, (b) for example, for $\eta = 0.99$ and $\kappa = 0.001$, and (c) for $\eta = 0.99$ and $\kappa = 0.0015$; other parameters are as in Figure 3.4.

Figure 3.11 shows what this perturbation statement means on the level of the leading eigenvalues of the associated ECM as continued with DDE-BIFTOOL. In Figure 3.11a–c, we plot the real parts of the eigenvalues as a function of the DDE-BIFTOOL point number of the computed points in a continuation in the feedback phase C_p (for fixed κ). Along gray parts, the respective eigenvalue is real, while black curves correspond to a pair of complex-conjugate eigenvalues. Figure 3.11a shows the continuation of the leading eigenvalues of the limiting and degenerate case of $\eta = 1.0$. There are two branches of real eigenvalues that cross the zero axis at the two double-zero points DZ. Notice further that there is a real eigenvalue zero independent of C_p (or the point number). We find that, in this (nondimensionalized) representation, the eigenvalue spectrum for $\eta = 1.0$ does not depend on κ, as long as the curve DZ in Figure 3.10f is indeed crossed (only twice). Figure 3.11b shows the perturbation of the leading eigenvalues for $\eta \approx 1.0$ for the case that the bifurcation diagram in Figure 3.10e is crossed horizontally below SH_3, for example, for fixed $\kappa = 0.001$. The ECM is stable for large and small point numbers. It is destabilized (for increasing or decreasing point number or C_p) in the supercritical Hopf bifurcation H_3, which takes place just before the saddle-node bifurcation S. The associated pair of complex eigenvalues becomes real below and above H_3. Figure 3.11c shows the perturbation of the leading eigenvalues for $\eta \approx 1.0$, for the case that the bifurcation diagram in Figure 3.10e is crossed above SH_3, for example, for fixed $\kappa = 0.0015$. The ECM is still stable for large and small point numbers, but this time it is destabilized (for increasing or decreasing point number or C_p) in the supercritical saddle-node bifurcation S, which takes place just before H_3. Again, the pair of complex eigenvalues becomes real below and above H_3. Overall, Figure 3.11 shows that in both cases the limit for $\eta = 1.0$ is reached as follows: the respective bifurcation points S and H_3 and the nearby points where two eigenvalues become complex converge to the point DZ. At the same time, the three branches in Figure 3.11a are approached by corresponding branches in Figure 3.11b and c.

3.5
Stability and Bifurcations of Periodic Solutions

The changes of the ECM bifurcation diagram with the homotopy parameter η have immediate consequences for bifurcating periodic solutions. As was already mentioned, there are two different types of periodic intensity fluctuations. First, undamped relaxation oscillations that correspond to a periodic exchange of energy between the electric field and the inversion at a characteristic RO frequency of around 5 GHz; ROs are present in all semiconductor lasers, and they can easily be undamped by external optical influences. Second, external round-trip oscillations, which are due to the travel of the light to the mirror and back to the laser. Hence, EOs generally have a much lower frequency – of around 500 MHz for the delay time $\tau = 500$ considered here. Both types of oscillations arise by crossing supercritical parts of Hopf bifurcation curves in the bifurcation diagrams of Figures 3.4 and 3.6. The question arises which branches of Hopf bifurcations give rise to ROs and EOs, respectively, and how large their respective stability regions are.

As Figure 3.12 shows, the answer depends strongly on the homotopy parameter η. Shown are two-parameter bifurcation diagrams of the periodic solutions of Equations 3.12–3.14 in the (C_p, κ)-plane for four values of η. Specifically, we determined the stability regions of ROs and EOs by computing bifurcation curves of periodic solutions with the package PDDE-CONT [18], which is able to deal with the additional S^1-symmetry of the DDE. Apart from saddle-node and Hopf bifurcation curves, we also found curves of period-doubling bifurcations PD, of torus (or Neimark–Sacker) bifurcations T, and of homoclinic bifurcations hom. For simplicity, we show one instance of the relevant part of the bifurcation diagrams in Figure 3.12 over several periods of C_p.

For the special case of $\eta = 0$ in Figure 3.12a, there is a stability region of ROs that is bounded by two torus bifurcation curves T_2 and T_3, which emanate from the codimension-two points $DH_{1,2}$ and SH_1, respectively. We also find a very small region of EOs, which is bounded by the Hopf bifurcation curve H_2 and a torus bifurcation curve T_1. The latter curve emanates from the codimension-two saddle-node Hopf point SH_2 and closely follows H_2. Note that all bifurcation curves are shown twice in Figure 3.12a due to the additional π symmetry for $\eta = 0$. As was shown in Figure 3.7, when η is increased from zero, the double-Hopf point $DH_{1,2}$ moves rapidly up and to the left, leaving our region of interest. As it does so, it "drags" the torus curve with it, so that the RO stability region rapidly grows in size and is bounded by the Hopf bifurcation curve H_1 below and torus bifurcation curves T_2 and T_3 above. The shape and size of the RO stability region remain virtually unchanged when η is increased further (see Figure 3.12b–d). The situation is quite different for the EOs. The local change with η near the codimension-two points shown in Figure 3.7, in combination with the curve H_3 moving toward lower values of κ, results for moderate amounts of cross-feedback in an EO stability region of considerable size (see Figure 3.12b for $\eta = 0.3$). This region is bounded below by the supercritical part of the Hopf bifurcation curve H_3 and above by a curve of period-doubling and a curve of homoclinic bifurcations (that emerges from BT_3). Notice that stable EOs occur for

3.5 Stability and Bifurcations of Periodic Solutions

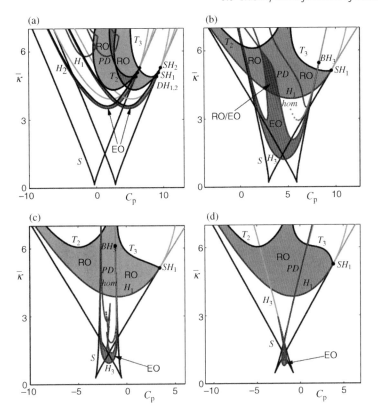

Figure 3.12 The two-parameter bifurcation diagrams of Equations in the (C_p, κ)-plane with the stability regions of periodic solutions. Shown are curves of saddle-node bifurcation S, Hopf bifurcation H_i, period-doubling bifurcation PD, torus bifurcations (T_i), and homoclinic bifurcation *hom*; regions of stable ROs are shaded light grey and regions of stable EOs are shaded dark grey; even darker shading illustrates overlap regions of coexisting stable periodic solutions. From (a) to (d), η takes the values 0.0, 0.3, 0.5, and 0.7; other parameters are as in Figure 3.4.

much lower values of κ compared to ROs. Furthermore, we find a large region of bistability between EOs and ROs for $\eta = 0.3$. As we have seen in Section 3.4.1, with increasing η the supercritical part of H_3 shrinks and moves to lower values of κ. It turns out that the curves of period-doubling and homoclinic bifurcations follow H_3, meaning that the EO stability region decreases dramatically as a result (see Figure 3.12c). As η approaches $\eta = 1.0$, this region effectively disappears near the lowest cusp point of the curve S (see Figure 3.12d).

Our results indicate that an experimental observation of EOs might be a practical way of testing for the existence of cross-coupling between the two transverse modes. Once EOs are found, a more ambitious goal would be to actually map out experimentally the size of the EO stability region as a function of both κ and C_p. It should be quite straightforward to distinguish EOs from ROs, as EOs have a much lower

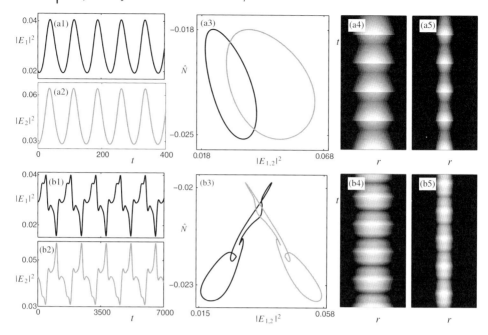

Figure 3.13 Coexisting relaxation oscillations in (a), and external round-trip oscillations in (b). Shown are time series (column 1) of the powers P_1 and P_2 of LP_{01} and LP_{02}, their projections into $(E_i, \hat{N})$-space (column 2), and the time series of the spatial intensity modes LP_{01} and LP_{02} of the laser (columns 3 and 4); $\eta = 0.3$, $C_p = \pi$ and $\kappa = 0.005$; other parameters are as in Figure 3.4.

frequency (which also makes it easier to measure them [37]). However, there is another distinguishing feature, which is illustrated in Figure 3.13 with specific examples of simultaneously stable ROs and EOs from the bistable region for $\eta = 0.3$ in Figure 3.12b. ROs are characterized by fast oscillations, where the two electric fields with amplitudes P_1 and P_2 are in phase (see Figure 3.13a1 and a2). In contrast, EOs are slower oscillations (notice the difference in scale of the t-axes), where the two electric fields with amplitudes P_1 and P_2 are in anti-phase (see Figure 3.13b1 and b2). A physical interpretation of the anti-phase behavior of the EOs might lie in the fact that the modes LP_{01} and LP_{02} compete for the spatial carrier source on a timescale that agrees with the diffusion time: one mode exhausts the carriers and then decays; while the carriers diffuse back, the other mode has a slightly larger overlap with the replenishing carrier profile, so that it starts lasing first, and the process repeats. We remark that on the level of the total spatial intensity of the laser, the individual maxima and minima of P_1 and P_2 add up for ROs, but they cancel each other out for EOs, leading to a less pronounced oscillation of the total laser power.

Figure 3.12 shows that ROs and EOs destabilize, when the feedback strength κ is increased. Two examples in Ref. [15] illustrate that this may lead to more complicated dynamics of the two modes involved: motion on an invariant torus bifurcating from ROs, and chaotic spatial mode dynamics after a period-doubling sequence starting from stable EOs. In both cases, the respective in-phase or anti-phase dynamics was shown to be preserved. The conclusion is that in certain regions of the (C_p, κ)-plane the system produces spatiotemporal chaotic dynamics due to the underlying interplay of temporal feedback with the spatial extent of the VCSEL.

3.6 Discussion

When one wants to model and study the dynamics of transverse spatial optical modes, that is, patterns of light, in a laser subject to optical feedback, two ingredients are crucial. First, one needs a mathematical description of the interaction of the spatial modes via their common spatial carrier reservoir that is provided by electric pumping and carrier diffusion. We considered here the case of a vertical-cavity surface-emitting laser with cylindrical geometry and a disk-shaped aperture. Starting from a spatial rate equation model that has been used in Ref. [11], we used a mode expansion in linearly polarized LP modes of a cylindrical waveguide. After discretization of the spatial carrier reservoir by Fourier modes, one obtains a system of ODEs that describes the dynamics of the LP modes (up to a given order and for both polarizations) inside the disk-shaped active region. Second, the feedback enters into the equation of the electric field of each mode, and it should be a function of the delayed electric fields of all modes. It is a major open question how each mode is influenced by itself and by the other modes as a result of the optical feedback from an external mirror. Note that in all previous studies pure self-feedback of the modes had been assumed. However, at least some cross-feedback between the modes via the feedback loop is likely to occur. Therefore, we model cross-feedback between modes by the introduction of additional homotopy parameters. This setting is still very general: the overall model is a (possibly quite large) DDE with (possibly very many) homotopy parameters in the feedback terms.

However, if one considers a VCSEL with only a few modes then the resulting DDE model may be small enough to allow for a bifurcation analysis with advanced numerical tools. In particular, the effect of self-feedback versus cross-feedback of the different modes (as modeled by the homotopy parameters) can then be investigated. We studied here as a concrete example a VCSEL with COF that supports only the first two rotationally symmetric LP modes; the expanded system of ODE has dimension 18, and a single homotopy parameter η suffices to describe the amount of self-feedback versus cross-feedback. The resulting DDE system is just small enough for numerical continuation tools and, in particular, it allows for the computation of the eigenvalue spectrum of periodic solutions. This allowed us to investigate how the stability regions of the external-cavity modes and bifurcating periodic solutions, in the plane of feedback phase C_p and feedback strength κ, depend on the homotopy parameter η – all the way from pure self-feedback to pure cross-feedback. We found a consistent bifurcation scenario that involves bifurcations of higher codimension, which, in effect, constitutes a three-parameter study of the stability properties of the two-mode VCSEL.

Our bifurcation analysis revealed specific results that are of interest from the physical point of view. First, we found large regions of coexistence between stable ECMs, and this multistability might be useful for application, for example, in all-optical flip-flop schemes. Moreover, we identified the existence of two distinct types of periodic solutions. The characteristic relaxation oscillations can be found for sufficiently large feedback strength κ for any value of η. However, we also found external round-trip oscillations, which show anti-phase dynamics of the two spatial modes on the timescale of the delay time. Importantly, EOs occur stably (in physically accessible regions of the (C_p, κ)-plane) only for low to intermediate values of η, that is, for moderate amounts of cross-feedback. Therefore, the bifurcation analysis suggests that in a two-mode VCSEL an experimental observation of EOs may be used as an indication of the amount of cross-coupling via the feedback loop.

An obvious challenge would be the study of VCSELs with general (not rotationally symmetric) spatial modes. In this case, the polarization of the modes will also play a role. The resulting DDE models will necessarily be quite a lot more involved, but the bifurcation analysis of ECMs in the presence of feedback may be feasible for a small number of spatial modes. Furthermore, the experimental verification of model predictions of VCSEL dynamics remains a considerable challenge. However, recent experiments in Ref. [38] have shown that individual spatial modes of a VCSEL can be stabilized via polarization- and frequency-selective feedback. Furthermore, high-speed imaging techniques as described in Ref. [37] might be a way of measuring oscillatory mode dynamics.

Acknowledgments

We would like to thank our long-term collaborator and coinvestigator Daan Lenstra for his valuable input. The work of Kirk Green was supported under the *Dynamics of Patterns Programme* of the Netherlands Organization for Scientific Research (NWO) and the Foundation for Fundamental Research on Matter (FOM).

References

1 Kane, D.M. and Shore, K.A. (2005) *Unlocking Dynamical Diversity: Optical Feedback Effects on Semiconductor Lasers*, Wiley-VCH Verlag GmbH, Weinheim.

2 Krauskopf, B. and Lenstra, D. (2000) *Fundamental Issues of Nonlinear Laser Dynamics*, AIP Conference Proceedings, vol. 548, American Institute of Physics Publishing, Melville, NY.

3 Fischer, I., Heil, T., and Elsäßer, W. (2005) Emission dynamics of semiconductor lasers subject to delayed optical feedback: an experimentalists perspective, in *Fundamental Issues of Nonlinear Laser Dynamics* (eds B. Krauskopf and D. Lenstra), American Institute of Physics Publishing, Melville, NY, pp. 218–237.

4 Lenstra, D., Verbeek, B.H., and den Boef, A.J. (1985) Coherence collapse in single-mode semiconductor lasers due to optical feedback. *IEEE J. Quantum Electron.*, **21**, 674–679.

5 Lang, R. and Kobayashi, K. (1980) External optical feedback effects on semiconductor injection laser properties. *IEEE J. Quantum Electron.*, **16** (3), 347–355.

6 Heil, T., Fischer, I., Elsäßer, W., Krauskopf, B., Green, K., and Gavrielides, A. (2003) Delay dynamics of semiconductor lasers with short external cavities: bifurcation scenarios and mechanisms. *Phys. Rev. E*, **67**, 066214.

7 Hess, O. (2000) Theory and simulation of spatially extended semiconductor lasers, in *Fundamental Issues of Nonlinear Laser Dynamics*, AIP Conference Proceedings, vol. 548 (eds B. Krauskopf and D. Lenstra), American Institute of Physics Publishing, Melville, NY, pp. 128–148, 173-L 190.

8 McInerney, J.G., O'Brian, P., Skovgaard, P., Mullane, M., Houlihan, J., O'Neil, E., Moloney, J.V., and Indik, R.A. (2000) High brightness semiconductor lasers with reduced filamentation, in *Fundamental Issues of Nonlinear Laser Dynamics*, AIP Conference Proceedings, vol. 548 (eds B. Krauskopf and D. Lenstra), American Institute of Physics Publishing, Melville, NY, pp. 173–190.

9 Degen, C., Krauskopf, B., Jennemann, G., Fischer, I., and Elsäßer, W. (2000) Polarization selective symmetry breaking in the near-fields of vertical cavity surface emitting lasers. *J. Opt. B: Quantum Semiclass. Opt.*, **2**, 517–525.

10 Gorman, M., El-Hamdi, M., and Robbins, K.A. (1994) Experimental observation of ordered states of cellular flames. *Combust. Sci. Technol.*, **99**, 37–45.

11 Valle, A. (1998) Selection and modulation of high-order transverse modes in vertical-cavity surface-emitting lasers. *IEEE J. Quantum Electron.*, **34**, 1924–1932.

12 Postlethwaite, C.M. and Silber, M. (2007) Spatial and temporal feedback control of traveling wave solutions of the two-dimensional complex Ginzburg–Landau equation. *Physica D*, **236**, 65–74.

13 Topaz, C.M., Porter, J., and Silber, M. (2004) Multifrequency control of Faraday wave patterns. *Phys. Rev. E*, **70**, 066206.

14 Kyrychko, Y.N., Hogan, S.J., Gonzalez-Buelga, A., and Wagg, D.J. (2007) Modelling real-time dynamic substructuring using partial delay differential equations. *Proc. R. Soc. A*, **463**, 1509–1523.

15 Green, K., Krauskopf, B., Marten, F., and Lenstra, D. (2008) Bifurcation analysis of a spatially extended laser with optical feedback. *SIAM J. Appl. Dyn. Syst.*, **8**, 222–252.

16 Hale, J.K. and Verduyn Lunel, S.M. (1993) *Introduction to Functional Differential Equations*, Springer, New York.

17 Engelborghs, K., Luzyanina, T., Samaey, G., and Roose, D. (2001) DDE-BIFTOOL: a Matlab package for bifurcation analysis of delay differential equations, Technical Report TW-330, Department of Computer Science, K.U. Leuven, Belgium, http://www.cs.kuleuven.ac.be/cwis/research/twr/research/software/delay/ddebiftool.shtml.

18 Szalai, R. (2005) PDDE-CONT: a continuation and bifurcation software for delay-differential equations, Budapest University of Technology and Economics, Hungary, http://www.mm.bme.hu/szalai/pdde.

19 Krauskopf, B. (2005) Bifurcation analysis of lasers with delay, in *Unlocking Dynamical Diversity: Optical Feedback Effects on Semiconductor Lasers* (eds D.M. Kane and K.A. Shore), Wiley-VCH Verlag GmbH, Weinheim, pp. 147–183.

20 Roose, D. and Szalai, R. (2007) Continuation and bifurcation analysis of delay differential equations, in *Numerical Continuation Methods for Dynamical Systems: Path Following and Boundary Value Problems* (eds B. Krauskopf, H.M. Osinga, and J. Galán-Vioque), Springer, New York, pp. 359–399.

21 Krauskopf, B., Van Tartwijk, G.H.M., and Gray, G.R. (2000) Symmetry properties of lasers subject to optical feedback. *Opt. Commun.*, **177**, 347.

22 Green, K. (2009) Stability near threshold in a semiconductor laser subject to optical feedback: a bifurcation analysis of the Lang–Kobayashi equations. *Phys. Rev. E*, **79**, 036210.

23 Olesen, H., Osmundsen, J.H., and Tromborg, B. (1986) Nonlinear dynamics and spectral behavior for an external cavity laser. *IEEE J. Quantum Electron.*, **22** (6), 762–773.

24 Rottschäfer, V. and Krauskopf, B. (2007) The ECM-backbone of the Lang–Kobayashi equations: a geometric

picture. *Int. J. Bifurcat. Chaos*, **17** (5), 1575–1588.

25 Erneux, T., Rogister, F., Gavrielides, A., and Kovanis, V. (2000) Bifurcation to mixed external cavity mode solutions for semiconductor lasers subject to optical feedback. *Opt. Commun.*, **183**, 467–477.

26 Haegeman, B., Engelborghs, K., Roose, D., Pieroux, D., and Erneux, T. (2002) Stability and rupture of bifurcation bridges in semiconductor lasers subject to optical feedback. *Phys. Rev. E*, **66**, 046216.

27 Guckenheimer, J. and Holmes, P. (1983) *Nonlinear Oscillations, Dynamical Systems, and Bifurcations of Vector Fields*, Springer, New York.

28 Kuznetsov, Yu. (2004) *Elements of Applied Bifurcation Theory*, 3rd edn, Springer, New York.

29 Yu, S.F. (2003) *Analysis and Design of Vertical Cavity Surface Emitting Lasers*, Wiley-VCH Verlag GmbH, Weinheim.

30 Dellunde, J., Valle, A., and Shore, K.A. (1996) Transverse-mode selection in external-cavity vertical-cavity surface-emitting laser diodes. *J. Opt. Soc. Am. B*, **13**, 2477–2483.

31 Green, K., Krauskopf, B., and Lenstra, D. (2007) External cavity mode structure of a two-mode VCSEL subject to optical feedback. *Opt. Commun.*, **277**, 359–371.

32 Onishi, Y., Nishiyama, N., Caneau, C., Koyama, F., and Zah, C.-E. (2006) All-optical regeneration using transverse mode switching in long-wavelength vertical-cavity surface-emitting lasers. *Jpn. J. Appl. Phys.*, **45**, 467–468.

33 Torre, M.S., Masoller, C., and Mandel, P. (2002) Transverse-mode dynamics in vertical-cavity surface-emitting lasers with optical feedback. *Phys. Rev. A*, **66**, 053817.

34 Tanguy, Y., Ackemann, T., and Jäger, R. (2006) Characteristics of bistable localized emission states in broad-area vertical-cavity surface-emitting lasers with frequency-selective feedback. *Phys. Rev. A*, **74**, 053824.

35 Erzgräber, H., Krauskopf, B., and Lenstra, D. (2006) Compound laser modes of mutually delay-coupled lasers. *SIAM J. Appl. Dyn. Syst.*, **5**, 30–65.

36 Fowler, D.H. and Thom, R. (1989) *Structural Stability and Morphogenesis: An Outline of a General Theory of Models*, Westview Press, Boulder, CO.

37 Barchanski, A., Genstry, T., Degen, C., Fischer, I., and Elsäßer, W. (2003) Picosecond emission dynamics of vertical-cavity surface-emitting lasers: spatial, spectral, and polarization-resolved characterization. *IEEE J. Quantum Electron.*, **39**, 850–858.

38 Chembo, Y.K., Mandre, S.M., Fischer, I., Elsäßer, W., and Colet, P. (2009) Controlling the emission properties of multimode vertical-cavity surface-emitting lasers via polarization- and frequency-selective feedback. *Phys. Rev. A*, **79**, 013817.

4
Optical Delay Dynamics and Its Applications
Laurent Larger and Ingo Fischer

4.1
Introduction

Dynamical systems, including optical systems, can often be described by coupled oscillators. The concept of coupled oscillators is indeed one of the most successful paradigms in engineering and science. In many real systems, the coupling is, however, not necessarily instantaneous. It might exhibit propagation delays that are comparable to the characteristic oscillatory timescales of the individual oscillators or even larger. The relevance of such delays in coupling or self-coupling (feedback) has been recognized and studied in systems like molecular oscillators [1], electronic circuits [2], Van der Pol oscillators [3], traffic models [4] and in generic model systems like phase oscillators [5] or logistic maps [6]. In the early days, however, only the situations of small delays and weak nonlinearities could be tackled. Optical systems have boosted the studies of delay systems, complemented by the development of novel analytical [7] and numerical tools [8, 9] for delay differential equations. Meanwhile, a multidisciplinary field of delay dynamics, its synchronization and its applications, has emerged (Chapter 3) [10, 11]. But why are delay systems so particular? The reason lies in the fact that the phase space of delay systems is an infinite-dimensional space of continuous functions C on the interval $[-\tau, 0]$.

In optical systems, delay effects on dynamical phenomena are often occurring generically and are technologically easier to obtain. Mainly two features are essential for the occurrence of intrinsic delays or for their easier realization in experiments:

- Propagation of light is characterized by the fastest known velocity. Purely delaying nearly any amplitude or phase variations of a light wave can be simply achieved via a short propagation distance in a medium. Usual transparent media are moreover only weakly dispersive, thus resulting in a nearly constant delay for a wide range of timescales, from long to very short ones.

- Light–matter interactions (dynamical effects) can occur on very short timescales, thus implying that even delays originating from very short propagation lengths cannot be neglected with respect to the short timescales of the optical (or electro-optic) dynamical system.

That optics was not the first field for the investigation of delay dynamical phenomena is partly due to the fact that it took technologically some time before the corresponding very short timescales of the light–matter interactions dynamics became accessible. For nearly one or two decades, we do have real-time GHz speed oscilloscopes, or 100 GHz streak cameras, which turned out to be key tools to get deeper into the experimental investigations of delay dynamical phenomena.

But it has been electromagnetic waves that, besides the above-mentioned examples, strongly stimulated interests and research efforts on delay dynamical effects already in the late 1950s, when the human space adventure started. The fastest dynamical phenomena involved in this context were much slower than the ultrafast light–matter interaction timescales: they were simply those of conventional "slow" electronics, with approximately microsecond timescales. However, the microwave frequencies used to carry the signal from earth to space were traveling on such long distances that delays involved in the dynamics of the earth–satellite distant control systems (a spatially spread dynamics) were strongly dominating the stability issues and the global dynamical behavior.

The giant progress made technologically to improve the speed of acquisition and instrumentation systems fortunately also allowed to observe the optical delay dynamical effects occurring on very short timescales, in the small space of a laboratory, but more comfortable and less expensive than in an earth–space experiment. More specifically, the development of optical telecommunications did an essential contribution to the fundamental investigation of delay dynamical phenomena in optics, through the design of ultrafast electronic, optoelectronic, and electro-optic devices.

This chapter will focus on delay dynamical systems in optics. We will discuss two different, but complementary, experimental approaches, both of which were prominent in the investigation of optical delay systems. One is typically referred to as the all-optical delay dynamics' approach. It consists, in its simplest implementation, of a semiconductor laser subjected to optical feedback from an external mirror. The laser is acting as nonlinear oscillator, described by nonlinear laser rate equations. The mirror introduces the feedback, described by a linear delay term in the dynamical equations [12]. Figure 4.1 depicts this concept. This can be extended to delay-coupled lasers with or without additional feedback from mirrors. The second approach, although originally inspired from an all-optical system (the Ikeda ring cavity [13]), was experimentally investigated through optoelectronic and/or electro-optic setups [14, 15]. The dynamical process of concern is ruled by a linear filtering in the electronic domain, whereas a delayed nonlinear term is involved as a drive of the linear filter, through a delayed nonlinear feedback loop oscillator architecture (Figure 4.1b).

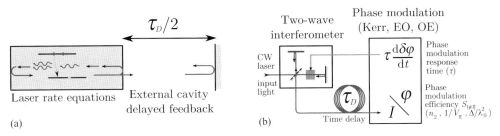

Figure 4.1 Principles of delay dynamics in optics. (a) External cavity laser diode. (b) Ikeda setup principle.

4.2
Experimental Setups

4.2.1
Delayed Feedback and Delay-Coupled Semiconductor Lasers

The first studies of delayed all-optical feedback on semiconductor lasers date back to the early 1970s [16]. It has already been recognized in these early investigations that optical feedback from external cavities can induce dynamical instabilities in semiconductor lasers, resulting in irregular intensity fluctuations. Meanwhile, it is known that the origin of these instabilities is the nonlinear interaction between the optical field in the laser and the delayed optical field due to the feedback. The reason that even external cavity lengths of only centimeters or millimeters results in relevant delays lies in the fast timescales of the interaction of optical fields with the charge carrier reservoir of the semiconductor medium. The relaxation oscillations of semiconductor lasers, representing the characteristic oscillations between photon reservoir and charge carrier reservoir, have typical pump-dependent periods of only 1 ns down to 50 ps. These timescales correspond to propagation lengths of light in the order of centimeters, thus making even delays of short external cavities nonnegligible. The nonlinearity of the interaction among the optical fields has its origin in the properties of the semiconductor gain medium. Due to different phenomena related to the band structure of the semiconductor medium, like the asymmetry of the gain spectrum, and due to many-particle interactions, amplitude and phase of the optical field are coupled with each other. This amplitude-phase coupling results in a multitude of characteristic phenomena in semiconductor lasers, including their broadened linewidth, frequency chirp under intensity modulation, and last but not least in a number of dynamical instabilities.

The study of external cavity semiconductor lasers has resulted in a large number of publications. Although the motivation in the early days was more focused on the understanding of these instabilities in order to avoid or suppress them, a research field has emerged studying these systems as test beds for dynamical systems with delay. Finally, for a bit more than a decade, the application potential of these systems has been recognized. Since then, they are being studied with the motivation to utilize the dynamical instabilities – as discussed later – for example, for communication purposes, and tailored light sources.

The real experimental setups to study external cavity semiconductor lasers are a bit more demanding than the sketch in Figure 4.1a and exist in different variations. The external cavities can be integrated on chip, free space, or fiber cavities. The corresponding cavity lengths therefore range from submillimeter (integrated) to many kilometers (fiber). In the following, we will concentrate on free-space cavities and free-space coupling configurations with typical propagation lengths of centimeters to few meters, since they have been studied most extensively. Setups to study delay dynamics using free-space cavities typically comprise a mirror whose roughness is less than 5 % of the optical wavelength. The length of the cavity has to be adjustable on subwavelength range in order to be able to control the phase of the feedback field. The light coming from the laser is usually collimated such that the feedback light is refocused back into the laser chip. The feedback strength is controlled via neutral density filters. For detection, the light needs to be analyzed optically via an optical spectrum analyzer to determine the emitting modes and their wavelengths and a fast photodetector to characterize the dynamics of the emitted light intensity. The electrical output of the photodiode is further analyzed via fast real-time oscilloscopes and electrical spectrum analyzers. Due to the fast timescales of the dynamics, this is demanding and only since the past few years does exist the oscilloscope capable of acquiring the dynamics up to the fast relaxation oscillation timescales.

4.2.2
Nonlinear Delayed Electro-Optic Feedback

As stated in the introduction, the electro-optic approach is an experimental solution initially proposed to perform the "Ikeda ring cavity" Gedanken experiment. The differential process is here issued from a linear filtering effect. This effect originates in the Ikeda ring cavity from the rate of change of a Kerr phenomenon, that is, the rate of change of the refraction index with respect to the intensity. This rate of change is extremely fast in many Kerr materials with subpicosecond or even femtosecond characteristic response time, thus leading to experimental difficulties in order to observe and record ultrafast and nonrepetitive dynamics such as chaotic solutions. Another important issue that is preventing from a tractable experiment in the case of the Ikeda setup is the rather low efficiency of the light–matter interaction, that is, the rather low value in general of the Kerr refractive index $n_2 = \Delta n / \Delta I$. For significant refractive index changes, rather high optical intensities are required, thus complicating the actual use of the setup for practical applications (such as chaos communication, as will be described later). In the case of an electro-optic realization, the light–matter interaction involving an optical intensity into optical phase shift conversion is somehow replaced by optical to electrical conversion (via a photodetector) and electrical to optical conversions (via the linear electro-optic effect). The resulting rates of change are definitively slower since it involves the timescales of electronic devices. However, the huge technological progress achieved for high-speed optical telecommunications still allows to have easy access to subnanosecond timescales, and even tens of picoseconds, for which a few tens of centimeters of light propagation medium is enough to generate significant delays (10 cm leads to 330 ps delay in vacuum, or 500 ps in optical glass fiber). The remaining action to finalize the nonlinear delay dynamics is the nonlinear process itself. The latter is performed in the Ikeda setup through the modulation of an interference condition, due to the optical phase change provided by the Kerr medium of length L, $\delta\phi = 2\pi(n_2 I)L/\lambda$. Exactly the same interference phenomenon is used in the electro-optic realizations, except that the phase change $\delta\phi$ is performed through the linear electro-optic Pockels effect [14, 15, 18], $\pi V/V_\pi$, where V_π^{-1} measures the electro-optic efficiency. In the case of an original optoelectronic version of the Ikeda setup, the phase interference condition is ruled by a wavelength change of a tunable semiconductor laser [20] (with an imbalanced interferometer having an optical path difference Δ, the phase interference condition is given by $2\pi\Delta/\lambda \simeq 2\pi\Delta/\lambda_0 - (2\pi\Delta/\lambda_0^2)\delta\lambda$, with $\delta\lambda$ being the wavelength deviation from the central wavelength λ_0). The detailed phase variations just described underline the importance of the accessible phase variation range on the degree of the actual nonlinearity involved in the dynamics. Considering a two-wave interference phenomenon, a phase variation of π around constructive or a destructive interference leads to a parabolic (second order) nonlinearity. Around a quadrature phase shift interference condition, the nonlinearity is cubic (third order), but the required phase variation should be greater (of the order of 2π) so that two extrema can be clearly scanned over the corresponding phase fluctuations. If one can achieve even larger phase deviations, up 3, 4, or 5π, the polynomial equivalent degree of the scanned nonlinearity can be as high as 4, 5, or 6. In real-world experiments, it is nearly impossible to have high enough Kerr efficiency to achieve a nicely parabolic effect in the delay dynamics. The electro-optic approach is in that sense clearly superior to the Kerr-based setup, since modern telecom integrated electro-optic modulators can have V_π as low as 2–3 V, thus allowing the scanning of at least 4π when the drive voltage is about 12 V peak-to-peak. The wavelength dynamics is here even better with more than seven tunable π phase shifts ($\Delta \simeq 1$ cm, and $\delta\lambda \simeq 1.5$ nm); however, the main drawback here is the maximum rate of change of tunable laser (slow dynamics), which is at least three orders of magnitude slower than the electro-optic modulators (fast dynamics). The efficiency of the intensity to phase conversion will be quantified by a factor $S_{[\phi/I]}$.

The next section will establish dynamical models, intended to describe as accurately as possible the experimental behavior. The models will be used to explain analytically, and to recover numerically, the observed dynamical solutions, which will be reported and described in the third section. The final section will develop a few applications performed on the basis of different dynamical features offered by delay dynamics in optical or optoelectronic systems, which will be related to highly complex and chaotic regimes or on the contrary to strongly stabilized regular ones.

4.3
Dynamics: Modeling, Numerics, and Experiments

4.3.1
The All-Optical Feedback System and the Lang–Kobayashi Rate Equations

The description of the dynamics of semiconductor lasers with delayed feedback requires a model that, as minimal requirement, needs to comprise a gain medium description taking into account the amplitude-phase coupling of the field as nonlinearity and a delayed coherent feedback term. While a number of models exist, which have been developed depending on the questions being addressed, the Lang–Kobayashi model has become very popular. It is a minimalistic model, yet capturing a large amount of the dynamical phenomena on a qualitative and often even quantitative level. The Lang–Kobayashi rate equations [12] are equations for the excess number of carriers $n(t) = N(t) - N_0$ with respect to the solitary threshold level N_0 and for the slowly varying complex electrical field amplitude $E(t)$:

$$\dot{E}(t) = \frac{1}{2}(1+i\alpha)g_n n(t) E(t) + \kappa E(t-\tau)e^{-i\omega_0 \tau}, \qquad (4.1)$$

$$\dot{n}(t) = (p-1)\frac{I_{\text{th}}}{e} - \gamma_e n(t) - (\Gamma_0 + g_n n(t))P(t). \qquad (4.2)$$

The delayed optical feedback is taken into account via the feedback term, including κ as feedback rate and τ as delay time. The optical field is normalized such that $P(t) = |E(t)|^2$ is the photon number; ω_0 represents the optical frequency of the solitary laser, g_n the differential gain, Γ_0 the cavity decay rate, γ_e the inverse carrier lifetime. I_{th} is the bias current at solitary laser threshold, e the electron charge, and p the pump parameter. From the analysis of the Lang–Kobayashi equations, it is known that the fixed points of the external cavity system are of two types: fixed points corresponding to constructive interference of the coupled cavities, usually referred to as *external cavity modes*, and fixed points (saddle points) corresponding to destructive interference of the coupled cavities and, thus, referred to as *antimodes*. Each fixed point of the system is defined by a constant optical frequency $(\omega_0 + \Delta\omega)$ and a constant carrier number n. Modes and antimodes lie on an ellipse around the solitary laser mode in the $(\Delta\omega, n)$ space [21].

A linear stability analysis of the fixed points of the system shows that the modes destabilize for increasing optical feedback via Hopf bifurcations. However, the so-called maximum gain mode on the very low-frequency high-gain side of the ellipse and typically a few more in its direct vicinity always remain stable [22]. The dynamical instabilities due to delayed optical feedback can be interpreted as the itinerancy of the trajectory among the destabilized external cavity modes toward the stable maximum gain mode [23]. Consequently, the average emitted power increases during this itinerancy, while on fast timescales the emission is organized in picosecond pulses [24]. Toward the maximum gain mode, the basin boundaries of the attractor ruin of modes and antimodes approach such that a collision (or crisis) can occur. The

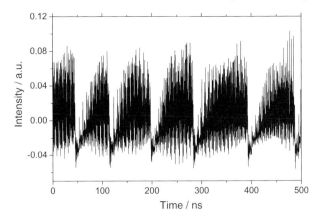

Figure 4.2 Delayed all-optical feedback dynamics of a semiconductor laser measured in the long-cavity regime. Detection bandwidth was 3 GHz.

characteristic power dropouts, referred to as low-frequency fluctuations (LFF), observed at low pump parameters are associated with such a crisis ejecting the trajectory toward the lower gain part of the ellipse and the process restarts. Thus, the feedback dynamics can be understood as a global dynamics involving many unstable fixed points and being characterized by a combined intensity and optical frequency dynamics. Figure 4.2 depicts a typical experimental time series of the intensity dynamics of a laser with delayed optical feedback in the LFF regime.

The different timescales involved, including the fast picosecond pulsations and slower power dropouts (LFF), can clearly be identified.

4.3.2
The All-Optically Delay-Coupled Lasers

4.3.2.1 Two Mutually Delay-Coupled Lasers: Generalized Synchronization and Symmetry Breaking

A natural extension of the investigations of lasers with delayed self-coupling is the study of mutually delay-coupled semiconductor lasers. Two longitudinally delay-coupled semiconductor lasers, as schematically depicted in Figure 4.3, represent the simplest possible configuration in this class of systems.

They can be considered a paradigm for delay-coupled relaxation oscillators in general. Nevertheless, it is only about a decade ago that dynamical properties considering a delay in the coupling of semiconductor lasers have been focused on. Hohl *et al.* studied the dynamics of two mutually coupled but nonidentical semiconductor lasers for weak coupling. They found that the coupled lasers can exhibit localized synchronization characterized by low-amplitude oscillations in one laser and large oscillations in the other. The laser intensities exhibited periodic or quasi-periodic oscillations. Instabilities resulting in a variety of chaotic behaviors have been found in 2001 by Heil *et al.* [25] and Fujino and Ohtsubo [26] for weak-to-moderate

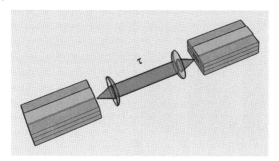

Figure 4.3 Two face-to-face delay-coupled semiconductor lasers.

coupling. Since these discoveries, interest in the fundamental investigation of mutually delay-coupled lasers has emerged, resulting in a considerable number of publications. One of the key criteria for the classification of the dynamical behavior turned out to be the coupling delay. Qualitatively different behavior has been found for the cases of short [27] and long delays [25]. If the coupling delay τ is of the same order as the relaxation oscillation period τ_{RO} of the laser ($\tau \sim \tau_{RO}$), we refer to it as the *short delay regime*. If $\tau \gg \tau_{RO}$, we denote it as the *long delay regime*. In the following, we concentrate on the long delay regime.

The long delay regime, defined by $\tau \gg \tau_{RO}$, is typically given for geometric coupling distances of $l > 30$ cm, corresponding to coupling delays of $\tau > 1$ ns. First we consider the completely symmetric situation: this includes the choice of very similar lasers, adjustment of their wavelengths, identical operating conditions, and symmetric bidirectional coupling. In Ref. [25] it has been found that the delayed coupling induces chaotic intensity dynamics on timescales ranging from subnanoseconds to microseconds. Figure 4.4 depicts a typical intensity time series of one of the lasers.

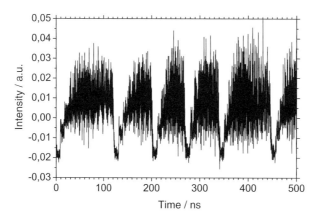

Figure 4.4 Intensity dynamics induced by delayed coupling under symmetric conditions.

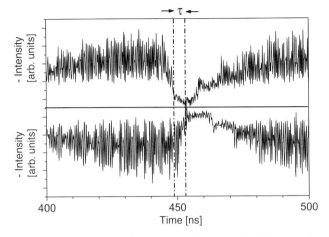

Figure 4.5 Comparison of the intensity dynamics of both lasers under symmetric conditions.

The dynamics resembles the dynamics found for delayed optical feedback. It shows similar low-frequency fluctuations, but here it does not originate from passive feedback, but from delayed coupling due to the respective other laser. As a next step, the intensity fluctuations of the two lasers have been compared with respect to each other. Figure 4.5 depicts the intensity dynamics of both lasers.

For ease of comparison, the second time series has been vertically flipped. One can see that the lasers show correlated power dropouts and even correlated sub nanosecond oscillations, however occurring not at the same time. The maximum correlation – reaching values of $C > 0.9$ – is obtained for a relative time shift. This time shift was found to be given roughly by τ or $-\tau$ [25]. Thus, although the configuration is completely symmetric, the behavior is not, in the sense that the lasers are not identically synchronized. They exhibit a form of generalized synchronization in which their behaviors are determined by the dynamics of the respective other laser, but their relation can be very complicated and often is not even known. The maxima in the cross-correlation function occurring at $+/-\tau$ indicate that one laser follows the respective other laser with a delay, however only showing similar, not identical, dynamics. Therefore, this type of generalized synchronization has also been referred to as leader–laggard-type synchronization. Although for completely symmetric conditions, leader and laggard role are spontaneously chosen and can change in time, they can also be controlled by introduction of slight asymmetries. One way to achieve this is by relative spectral detuning of the emission of the two lasers. However, for nominal detunings of only about 1 GHz that is small to typical optical locking ranges of larger than 10 GHz, the leader and laggards role can be fixed. For edge-emitting lasers, the laser with higher frequency becomes the leader in the dynamics [25, 28].

The emission dynamics of delay-coupled laser configurations – here in particular for two mutually delay-coupled lasers – can again be modeled (assuming single solitary laser mode emission and low-to-moderate coupling) via a set of rate equations, resembling the Lang–Kobayashi equations for the laser with feedback:

$$d_t E_{1,2}(t) = \mp i\Delta E_{1,2} + \frac{1}{2}(1-i\alpha)\left[g_{n,i}n_i\right] E_{1,2} + \kappa_c E_{2,1}(t-\tau), \tag{4.3}$$

with κ_c being the coupling strength. E represents the slowly varying electric fields around the symmetric reference frame $(\Omega_2 + \Omega_1)/2$, and Δ represents the detuning of the lasers in this reference frame. $(\Omega_{1,2})$ is the free-running optical frequency of laser 1 and 2, respectively. The complementary equations for the excess carrier densities n_i read as follows:

$$\dot{n}_i = (p-1)\frac{I_{\text{th},i}}{e} - \gamma_e n_i - (\Gamma_0 + g_{n,i}n_i(t))\|E_i\|^2, \tag{4.4}$$

where, as before, $E_i(t)$ refers to the optical field generated by laser i and $g_{n,i}$ the differential gain of laser i. $I_{\text{th},i}$ is the bias current at the solitary threshold of laser i, e is the electron charge, and p is the pump parameter. $\|\cdots\|$ denotes the amplitude of the complex field.

From the experimental studies, even if one argues that asymmetries in the setup or laser equations are the origin of the symmetry breaking, resulting in the observed leader–laggard behavior, in the modeling one can choose perfectly symmetric conditions. Still, the same leader–laggard behavior is being found, representing the generalized synchronization. Due to the symmetry in the system, a symmetric solution has to exist and in the modeling one can even prepare the system to start in this solution. However, as soon as one laser experiences a tiny perturbation, the system escapes to the generalized synchronized solution. The symmetric solution is unstable. This is shown in Figure 4.6.

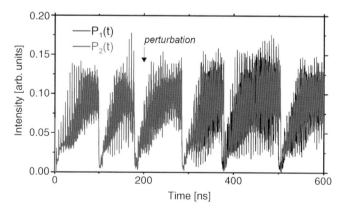

Figure 4.6 Numerically obtained intensity dynamics of two mutually delay-coupled lasers. At $t = 0$, the system is prepared in the isochronously synchronized state. At $t = 200$, a small perturbation is applied. As a consequence, the system emerges into the leader–laggard state. Courtesy of Claudio Mirasso.

Remarkably, this behavior that the symmetric, isochronously synchronized solution is unstable in delay-coupled oscillators holds not only for the chosen parameter conditions but also for all considered parameter situations and even for a large class of delay-coupled oscillators in general. It is only recently that we have began to understand why this is the case [29]. So, naturally, the question arose whether the zero-lag solution always has to be unstable, or whether this can be overcome by modifying the coupling configuration. This question resulted in the extension to a chain of three mutually delay-coupled semiconductor lasers, as discussed in the following section.

4.3.2.2 Delay-Coupled Lasers with a Relay: Zero-Lag Synchronization

Chain of Three Mutually Delay-Coupled Lasers The study of a chain of three mutually delay-coupled lasers as depicted in Figure 4.7 was motivated by the assumption that a relay in the middle between two identical lasers might stabilize an isochronously synchronized solution despite delay. This solution is also referred to as zero-lag synchronization. There had been indications that stable zero-lag synchronization of delay-coupled elements should, in principle, be possible. In neurophysiological experiments, Roelfsema *et al.* had found correlations at zero lag between different cortical areas, although these areas were connected only via long-range connections with significant delay [30]. Since synchronization is assumed to play an essential role for performing cognitive tasks like feature binding, the identification of topologies and mechanisms allowing zero-lag synchronization is an important problem. This problem is also addressed in Chapter 10.

As a start, numerical modeling has been performed based on a rate equation model in analogy to those discussed in the previous sections. Details of the model and chosen parameters can be found in Ref. [31]. We have chosen identical operating conditions, identical laser parameters, identical delays, and identical coupling strengths. From the numerics we obtained the emission dynamics, as depicted in Figure 4.8. They have been obtained for $I_1 = I_2 = I_3 = 17.5$ mA $\approx 1.15\%$ I_{th}. The figure shows time series of respective pairs of lasers in the left column, and the corresponding cross-correlation functions in the right column.

From Figure 4.8 one can see that indeed the two outer lasers exhibit identical synchronization at zero lag. Neighbors, meaning the center laser and one of the outer

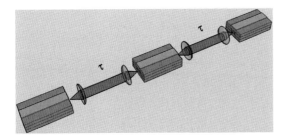

Figure 4.7 Scheme of laser chain with three mutually delay-coupled lasers.

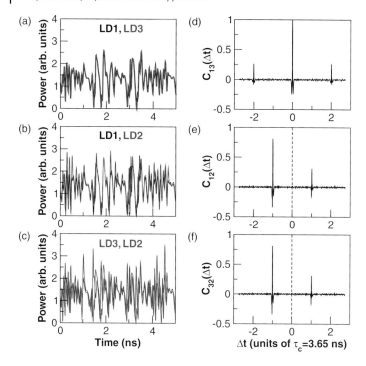

Figure 4.8 Numerical modeling results of laser chain with three mutually delay-coupled lasers. (a–c) Pairs of numerical time traces of the output intensity of the lasers in the low-frequency fluctuation regime and for the case of zero detuning between the three lasers. (d–f) Cross-correlation functions of the corresponding time series. The time series of the central laser have been shifted by τ_c to allow easier comparison. The bias current in the three lasers is 1.15% above threshold (from Ref. [31]).

lasers, respectively, show only generalized synchronization of the leader–laggard type. The corresponding cross-correlation functions for pairs of direct neighbors show that the center laser is even lagging behind the outer lasers, therefore not simply driving them. This underlines that the observed behavior is an emergent collective dynamics of this mutually coupled configuration.

The modeling results raised the question whether the zero-lag synchronization solution is sufficiently stable to be found in experiments, where differences in laser parameters, coupling strengths, and coupling delays are inevitable. Therefore, experiments were performed for this configuration.

Three Fabry–Perot lasers emitting at $\lambda = 655$ nm have been chosen, exhibiting typical parameter deviations of the order of a few percent. The central laser was bidirectionally coupled to two outer lasers by mutual injection of their lasing TE-polarized fields. The central laser, which did not need to be carefully matched to the other two, mediated their dynamics. The outer lasers were placed at the same distance of ~ 1.1 m from the central one, resulting in coupling times of $\tau \sim 3.65$ ns. To avoid influence from the nonlasing TM modes, a polarizer was placed before the input of laser 2. Balanced beam splitters were used to send fractions of the light to the other lasers and to the corresponding photodetectors. The coupling strength was adjusted using neutral density filter. The lasers were pumped slightly above their lasing thresholds ($I_{dc} = 1.07 I_{th}$) and the pump current and temperature were controlled with high-precision controllers.

Without coupling, the three lasers emitted constant output power. Due to the mutual interaction, the lasing threshold current of the lasers was reduced by 5–10%, indicating moderate coupling strength. The laser outputs were detected by a fast photodetector (12 GHz bandwidth) whose signal was recorded and analyzed by a 4 GHz oscilloscope. With the light blocked between the central laser and one of the outer lasers, the system reduced to the case of two mutually injected lasers, which were discussed in the previous section. When the blocking was removed, the three lasers also exhibited delayed coupling-induced chaotic emission dynamics. However, now we found both outer lasers to synchronize without any time lag. The central laser was either leading or lagging the outer lasers, depending on parameters.

Figure 4.9 shows the time series of the output intensities (left column), in pairs, and the corresponding cross-correlation functions in such a way that a maximal cross-correlation at a positive time difference Δt max indicates that element j is leading element i by the time Δt max, and vice versa. For optimal synchronization quality, the optical frequency of the central laser was adjusted via temperature tuning. Zero-lag synchronization between the intensities of the outer lasers can clearly be seen in Figure 4.9a and d, in time series and cross-correlation function. Despite the inevitable differences between the lasers, a maximum of 0.86 could be found in the cross-correlation function at $\Delta t_{max} = 0$ (i.e., at zero lag). The correlation between the central laser and the outer ones (Figure 4.9b and c) is not as high, and exhibits a nonzero time lag, as can be seen from the cross-correlation functions shown in Figure 4.9e and f, which yield maxima of 0.56 and 0.59, respectively. Moreover, the maximum of this cross-correlation occurs at $\Delta t_{max} = 3.65$ ns. This lag coincides with the coupling time between the lasers. The fact that Δt_{max} is negative means that the central laser dynamically lags the two outer lasers. Therefore, it can be excluded that the outer lasers are simply driven by the central one.

We also observed that zero-lag synchronization solution is quite robust against spectral detuning of the center laser exceeding 15 GHz with correlations larger than 0.8. When studying numerically the influence of other parameters mismatch, we found that the central laser can have large parameter mismatch without preventing the occurrence of zero-lag synchronization [32]. Analogously, the zero-lag synchronization is also robust for relative mismatch between the outer lasers, although the acceptable tolerances are smaller in this case. Applying external perturbations via the pump current did not destroy the synchronization. Even more, we have verified that zero-lag synchronization also maintains for pump current modulation of the lasers. In addition, identical synchronization persists even for mismatches of the coupling delay times between the lasers, resulting in a temporal offset corresponding to the difference of the coupling delay times.

Semitransparent Mirror as Relay Element A main argument, why zero-lag synchronization could be observed in the laser chain, was that the relay laser in the center was

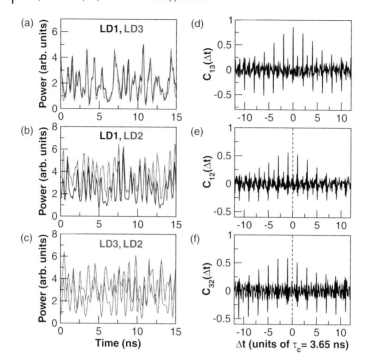

Figure 4.9 (a–c) Pairs of time traces of the output intensity of the lasers for the case of a central laser with slight negative detuning. (d–f) Cross-correlation functions of the corresponding time traces. The time series of the central laser have been shifted by τ_c to allow easier comparison (from [31]).

redistributing the signals of the respective outer lasers symmetrically. In addition, it was found that the parameters and operating conditions of the center laser were not crucial to observe zero-lag synchronization of the outer lasers. Consequently, the question arose whether the central laser could be replaced by a semitransparent mirror as relay element. The corresponding scheme is depicted in Figure 4.10.

The two semiconductor lasers are mutually coupled through a partially transparent mirror placed in the coupling path between both lasers. Therefore, the light injected into each laser is the sum of its delayed feedback from the mirror and the light coming from the respective other laser. For the numerical studies, coupling coefficients and

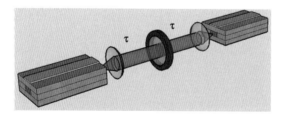

Figure 4.10 Scheme of two mutually delay-coupled lasers with semitransparent mirror as relay.

feedback strengths were chosen such that the lasers operate in a chaotic regime. In numerical investigations of this configuration, identical synchronization between the dynamics of both lasers was obtained for arbitrary coupling distances between the lasers. For the laser being precisely in the center, zero-lag synchronization was found. Changing the position of the mirror turned out not to be relevant for the synchronization quality. Even for strongly asymmetric positioning of the mirror, identical synchronization was still observed, with a temporal offset given by the difference of the corresponding delay times. Thus, identical and even zero-lag synchronization can be achieved with different realizations of the relay element. A parameter that, however, turned out to be critical for the semitransparent mirror configuration for obtaining good synchronization quality was phase differences between the optical coupling and feedback phases. Although the experiments in the all-optical scheme are difficult to control sufficiently, using electro-optic systems successful identical synchronization could be demonstrated experimentally [33].

Stability of Zero-Lag Synchronization   The results discussed above show that zero-lag synchronization can be achieved in delay-coupled lasers with a relay element in the center. However, it is not clear under which conditions this solution is stable, or whether it is even unconditionally stable due to the common driving of the outer lasers through the relay element. A detailed stability analysis showed the existence of unstable regimes, in particular for not sufficiently strong coupling [34]. In addition, it was found that even in large regimes where the synchronization manifold is transversely stable, characterized by negative transverse Lyapunov exponents, bubbling can still occur. Bubbling is the phenomenon of eventual escapes from a synchronization manifold due to an invariant set being transversely unstable. Responsible for the occurrence of bubbling are saddle points, corresponding to destructive interference conditions of the optical field in the outer lasers and the incoupled fields. These saddle points are not only crucial for the onset of the coupling-induced dynamical instabilities, as discussed in Section 4.2.1, but also for eventual escapes from the synchronization manifold, resulting in bubbling behavior.

Zero-Lag Synchronization in Neuronal Systems   All these results lead us back to the initial motivation to study and understand the zero-lag synchronization, the observation of it in neurophysiological experiments. Although the results above represent a mechanism allowing the achieving of zero-lag synchronization, it is not at all clear that this mechanism plays a role in the observed synchronization in the brain. Therefore, the relay scheme has been transferred to neuronal models. The investigations comprise models of individual neurons and of neuronal ensembles [35]. Indeed, it could be found that neurons and neuronal ensembles can be brought to zero-lag synchrony via a relay element, realized by individual neurons or another neuronal ensemble. In the brain, this relay could in some cases be realized by the thalamus. It should be noted that other explanations and configurations have been suggested to explain zero-lag long-range synchronization in the brain. Still, the mechanism of synchronization via a relay element represents an interesting candidate being robust over large parameter regimes.

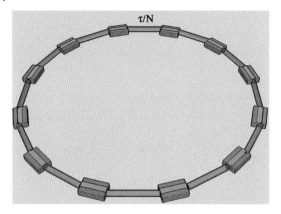

Figure 4.11 Scheme of a ring of unidirectionally delay-coupled lasers.

4.3.2.3 Rings of Unidirectionally Coupled Lasers with Delay

As networks of delay-coupled oscillators become more complex, it is crucial to understand the dynamics of its building blocks such as rings and arrays. Unidirectionally delay-coupled rings of oscillators have been studied in the context of traffic modeling [4] and in the field of gene expression, in particular synthetic genetic regulatory networks. As in the previous sections, we are interested in the dynamical behavior that emerges from the delayed coupling. Therefore, the solitary lasers again do not show chaotic dynamics when uncoupled. We were focusing on the question how the chaotic dynamics generated by a ring configuration of unidirectionally coupled lasers is changing with the number of composing elements keeping the total delay in the ring constant. More specifically, we were interested whether there is a relation between the chaotic dynamics generated by such rings of lasers and the dynamics of the previously discussed systems of a laser with delayed feedback and two delay-coupled lasers. Figure 4.11 shows a scheme of the unidirectionally delay-coupled lasers.

We consider a unidirectional ring configuration consisting of N identical semiconductor lasers placed in a ring of total delay τ. In the cases of $N = 1$ and $N = 2$, the ring configuration reduces to a single laser subject to delayed optical feedback and two mutually delay-coupled lasers, as depicted in Figure 4.1a and Figure 4.3. We have already discussed that the dynamics found in the case of $N = 2$ resembled the dynamics of a single delayed feedback laser. In how far the dynamics of these two cases and for arbitrary N can be quantitatively compared is being discussed in the following, based on numerical modeling. The modeling equations are an extension of the previously presented equations, taking into account the unidirectional coupling. For the complex optical field amplitude, they read as follows:

$$\dot{E}_j(t) = \frac{1}{2}(1 + i\alpha)g_{n,j}n_j(t)E_j(t) + \kappa E_{j-1}\left(t - \frac{\tau}{N}\right)e^{-i\omega_0\tau/N} \tag{4.5}$$

with $j = 0, 1, \ldots, N-1$. $g_{n,j}$ is the differential gain of laser j, ω is the frequency of the free-running laser, $\kappa = 40\,\mathrm{ns}^{-1}$ is the coupling coefficient, and $\tau = 1$ ns is the round-trip time in the ring. The following parameter values were considered: $\alpha = 5$, $g_n = 1.5 \times 10^{-8}\,\mathrm{ps}^{-1}$ for the differential gain, $\tau_{\mathrm{ph}} = 2$ ps for the photon lifetime, and $\tau_e = 2$ ns for the carrier lifetime. The pump current was chosen as $I = 1.5 I_{\mathrm{th}}$. For these parameters, the lasers are operating in the chaotic regime. Figure 4.12 shows an overview of the numerically obtained results for one SL with delayed optical feedback ($N = 1$), two mutually coupled lasers ($N = 2$), $N = 4$ and $N = 100$. In addition to time traces (left panel), power spectra (middle panel) and autocorrelation functions (right panel) are depicted to analyze and illustrate the changes in the dynamics.

Figure 4.12b shows the power spectrum for one laser with delayed optical feedback. The spectrum exhibits a broadband characteristic and on top the discrete frequency peaks related to the inverse delay time and multiples of it. Comparing the spectrum in Figure 4.12b with the power spectrum of bidirectionally coupled SLs depicted in Figure 4.12e, the peaks become less defined, while the underlying broadband spectral shape is unaffected. In Figure 4.12h, we find that this trend continues for $N = 4$. For $N = 100$, as shown in Figure 4.12k, the discrete peaks disappear completely. Although power spectra and autocorrelation functions are related via the Wiener–Khinchin theorem, the autocorrelation functions provide further insights into the characteristics of this behavior. For $N = 1$, the autocorrelation function exhibits several "delay echoes" separated by the total delay time τ, in addition to the short time correlations around $t = 0$. These echoes decrease with increasing N. Shape and amplitude of the peak around $t = 0$ remain unchanged when N is increased. All elements in the ring show identical power spectra and autocorrelation functions. This is regardless of the fact that the dynamics of different lasers can be uncorrelated with respect to each other. Analyzing the autocorrelation functions in more detail, it can be clearly seen that the correlation peak around $t = 2\tau$ in the case of one laser with delayed feedback $N = 1$ is precisely reproduced around $t = \tau$ in the autocorrelation function of the case $N = 2$. Also, the peak around $t = 2\tau$ for $N = 2$ is reproduced at $t = \tau$ for $N = 4$ and can also be found for $N = 1$ at $t = 4\tau$. A similar analysis can be performed for any peak. We found that shape and position of the correlation peaks are defined by the number of passes through a laser in the ring and not by the time it takes to complete those passes. Remarkably, one can exactly predict the autocorrelation function of a ring of N elements by selecting the corresponding peaks in the autocorrelation function of a single laser with delayed feedback. Certainly, this only holds as long as the response time of each element is smaller than the delay time, which guarantees that the correlations have decayed in between the peaks. But all the presented results have been in the regime of long delay times.

Altogether, these results on the unidirectional rings of delay-coupled oscillators connect all the previous sections of the all-optical scheme, showing how their dynamics is related.

4 Optical Delay Dynamics and Its Applications

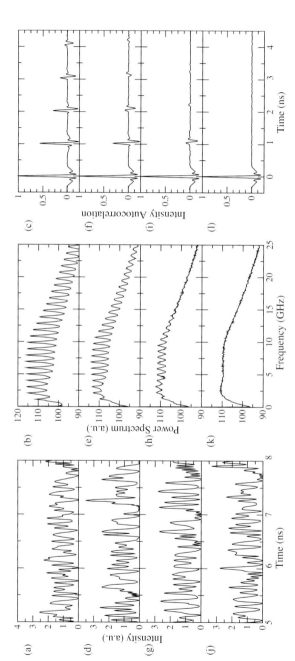

Figure 4.12 Time traces (left panels), power spectra (middle panels), and autocorrelation functions (right panels) for unidirectional rings of delay-coupled semiconductor lasers. (a–c) Semiconductor with delayed optical feedback ($N = 1$). (d–f) $N = 2$. (g–i) $N = 4$. (j–l) $N = 100$.

4.3.3
Linear Filtering of a Nonlinear Delayed Feedback

4.3.3.1 Modeling

The differential equation ruling an Ikeda-like delay dynamics can be derived according to the setup description in Figure 4.1b. The rate of change of the phase modulation is ruled in the frequency domain by a linear filtering. The original Ikeda dynamics assumes a first-order low-pass filter representing the rate of change for the Kerr effect. The notion of equivalent linear frequency filtering of a time domain rate of change is rather common in electrical engineering and signal theory. In physics, it is more common to represent the dynamics mostly in the time domain, thus resulting, for example, in the rate of change of atom populations according to the atomic level lifetime. The phenomenon is however similar, and from the signal processing point of view, the equivalent "low pass" dynamics of the Ikeda setup can be generalized to any other type of linear filter, for example, to bandpass type [36]. A different kind of filtering will in general imply different dynamics and bifurcation scenarios. The actual filtering corresponding to the electro-optic or optoelectronic setup usually originates from the combined bandwidth of the various electronic, optoelectronic, and electro-optic devices involved in the electronic path. Two kinds of filter families are then typically concerned with the electro-optical or optoelectronic delay dynamics found in the literature, a low pass kind characterized by a cutoff frequency $f_c = 1/(2\pi\tau)$ [13–15, 20], and a bandpass kind characterized by both a low and a high cutoff frequency, $f_l = 1/(2\pi\theta)$ and $f_h = 1/(2\pi\tau)$, respectively [17, 18, 36, 44]. The frequency filtering transfer functions of both kinds of filter are described in Equations 4.6 and 4.7. From the usual equivalence between the time and the Fourier domains (in which $\mathcal{I}(i\omega) = \mathrm{FT}[I(t)]$ and $\Phi(i\omega) = \mathrm{FT}[\phi(t)]$, and where we use $\mathrm{FT}^{-1}[i\omega\,\Phi(i\omega)] = (d/dt)\phi(t)$ and $\mathrm{FT}^{-1}[(i\omega)^{-1}\Phi(i\omega)] = \int \phi(t)dt$, a first-order delay differential equation is derived in Equation 4.6 for the first-order low-pass filtering, whereas an integro-differential one is obtained from the simplest second-order band-pass filter in Equation 4.7.

$$\frac{\Phi(i\omega)}{\mathcal{I}(i\omega)} = \frac{S_{[\phi/I]}}{1+i\omega\tau} \quad \xrightarrow{\mathrm{FT}^{-1}} \quad \tau\frac{d\phi}{dt}(t) + \phi(t) = S_{[\phi/I]}I(t), \tag{4.6}$$

$$\frac{\Phi(i\omega)}{\mathcal{I}(i\omega)} = S_{[\phi/I]}\frac{i\omega(\tau+\theta)}{(1+i\omega\theta)(1+i\omega\tau)} \quad \xrightarrow{\mathrm{FT}^{-1}}$$

$$\frac{1}{\theta}\int_{t_0}^{t}\phi(\xi)d\xi + \tau\frac{d\phi}{dt}(t) + \left(1+\frac{\tau}{\theta}\right)\phi(t) = \left(1+\frac{\tau}{\theta}\right)S_{[\phi/I]}(I(t)-I_0). \tag{4.7}$$

In the case of the delayed feedback architecture as depicted in Figure 4.1b, the drive input of the filter $I(t)$ is defined by the time-varying delayed interference state, which can be written as

$$I(t) = I_\varphi(t) = I_\beta \sin^2[\varphi(t-\tau_D) + \varphi_0], \tag{4.8}$$

where $\varphi = \phi/2$ was arbitrarily chosen so that the nonlinear function applies directly to the dynamical variable $\varphi(t)$. This nonlinear function $I(\varphi)$ thus strictly has a positive and a normalized unity amplitude, which is scaling that way the feedback gain of the delay dynamics. A usual normalized bifurcation parameter can be defined as $\beta = S_{[\varphi/I]} I_\beta$, where $S_{[\varphi/I]} = S_{[\phi/I]}/2$. This bifurcation parameter is simply tuned experimentally by adjusting the CW laser light intensity, as initially considered by Ikeda. Such a bifurcation parameter represents the gain of the delayed feedback loop oscillator or, more mathematically speaking, the weight of the nonlinear delayed feedback term (multiplicative amplitude of the normalized nonlinear function varying in the range $[0, 1]$). The value of β divided by π also gives a rough estimate of the polynomial order for the actual nonlinearity involved in the dynamics (two for quadratic, three for cubic, and so on).

The parameter φ_0 can also be used as a bifurcation parameter. Instead of a stretching of the nonlinear function as performed by β, it acts as a horizontal shift of it.

4.3.3.2 Delay Differential Dynamics

The fixed points of the delay differential dynamics in Equation 4.6 are simply the solutions of the equation $\varphi = \beta \sin^2(\varphi + \varphi_0)$, which can be graphically found as the intersections between the nonlinear feedback function $y = \beta \sin^2(\varphi + \varphi_0)$ and the first bisector $y = \varphi$. The stability analysis of these fixed points directly involves the slope of the nonlinear function at the considered fixed point, in which absolute slope value has to be smaller than 1 to ensure stability. The slope being proportional to β, increasing from zero the feedback gain usually leads to the destabilization of an initially stable fixed point. Compared to another well-known scalar (low-pass) delay differential equation, the Mackey–Glass dynamics [37], important differences have to be noticed: we do have a multiple extrema nonlinear function in the Ikeda dynamics, which is leading to more possible fixed points and to more complex dynamics. Variation of β or φ_0 easily leads to appearance or disappearance of fixed point, thus also contributing to more complex dynamics. It is also conjectured that the entropy of chaotic oscillations indeed increases linearly with β in the multiple extrema situation [38], whereas it saturates in the case of a single extrema Mackey–Glass dynamics. The first bifurcation of the stable fixed point however usually occurs for both cases in a Hopf bifurcation, after which a $2\tau_D$ period limit cycle appears (see details in Ref. [39] and references therein). In the large delay case ($\tau_D \gg \tau$), the limit cycle rapidly converges to a square waveform, which plateaus are found to be stable fixed points of the double iterated nonlinear function $I_\varphi \circ I_\varphi$ (thus also explaining the $2\tau_D$ periodicity of the cycle, which corresponds to twice round-trip in the delay oscillator loop). Further increasing β, a period-doubling bifurcation cascade is typically observed, similar to the one obtained for the corresponding Ikeda map $x_{n+1} = \beta \sin^2(x_n + \varphi_0)$ (where the iteration step form n to $n+1$ corresponds to the time delay τ_D). The limit from Equation 4.6 to the map, as $\tau/\tau_D \to 0$, is however singular, and the bifurcation sequence between the map and the delay differential dynamics can only be roughly compared (see a proposed continuous experimental transition scenario between the two dynamics in Ref. [40]).

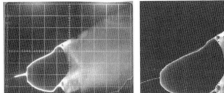

Figure 4.13 Bifurcation diagrams of the Ikeda dynamics (*left*: experiment; *middle*: numerics; *right*: numerics of the singular limit map model).

As an illustration, Figure 4.13 represents three different bifurcation diagrams of the low pass dynamics, when continuously increasing β (along the horizontal axis) and with $\varphi_0 \simeq 1.3$. The vertical axis reflects in gray scale (black, zero probability; white, high probability) the statistics of the transient-free dynamics for each value of β (statistics for a given dynamical regime are vertical cuts). The first bifurcation diagram is an experimental one, the second is obtained from the numerical integration of Equation 4.6, and the last corresponds to the map dynamics.

4.3.3.3 Integro-Differential Delay Dynamics

The low pass model is actually more attached to the fundamental studies of academic delay dynamics (such as the Ikeda or Mackey–Glass dynamics). The bandpass model is more motivated by technological and applications constraints, and it indeed appeared in the literature together with applications-oriented research, concerned either with pure tone microwave frequency generation [17, 41, 44] or with high-speed (thus broadband) optical chaos communications [42, 43]. Although these two applications have been performed by the same experimental architecture (Figure 4.14a), they are involving drastically different physical parameters. For pure tone microwave oscillators, a bandpass-selective (resonant) filter is required in the electronic feedback. The aim is to preferably select one single delay mode (the maximum gain mode in the Fourier domain, around 10 GHz, with typically

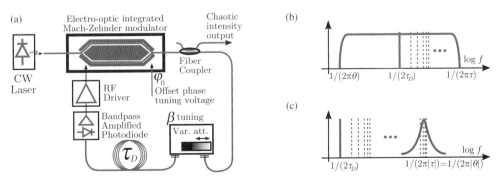

Figure 4.14 Bandpass electro-optic nonlinear delay dynamics. (a) Setup for the electro-optic delay bandpass oscillator. (b) Broadband filtering, cutoff frequencies, and delay modes. (c) Narrow filtering, cutoff frequencies, and delay modes.

10 MHz–3 dB bandwidth). Together with the selective RF filtering, a huge optical delay line is specifically used to fulfill the long storage time required for high spectral purity (a few kilometers of fiber, storing more than 10^5 periods of the microwave oscillation). For broadband chaos communications, the chaotic oscillation needs to spectrally hide a typical high bit rate signal spectrum that is extending from a few tens of kilohertz to tens of gigahertz: the feedback filter has to allow all these Fourier frequencies within the chaotic motion, and it is thus performed by an ultrawide bandpass amplifier. A wideband low-pass filter could also be suitable in principle, however it is technologically difficult to have DC-10 GHz electronic amplifier, thus leading to necessary bandpass filters. The delay line required to achieve high-complexity chaos only needs to be at least 10 times greater than the high cutoff response time τ (typically 12 ps for 13 GHz cutoff), and a few meters of fiber pigtails are practically far from enough for the delay (leading to a few tens of nanoseconds, thus $\tau_D/\tau \simeq 10^3$). The two applications will be reported in the final section. The interesting fundamental dynamical phenomena attached to each configuration (broadband or narrow feedback filter) will be described in the following section.

Broadband Bandpass Filter: Moderate Delay [18] The low-frequency cutoff is here related to the lowest frequency in the spectrum of a Gb/s information signal, a few tens of kilohertz. Practically, we have $\theta \simeq 5\,\mu s$, whereas $\tau \simeq 10\,ps$, the conditions that allow the spanning of the frequency components of the chaos over more than six orders of magnitude. The delay mode frequency $1/(2\tau_D)$ is in the particular experimental configuration, approximately set at the geometrical center of the bandpass filter (this is an arbitrary setting in some sense, since only $\tau_D/\tau \gg 1$ is practically required for a high-complexity chaotic attractor). As a consequence, more than thousands of delay modes can be mixed within the spectrum of the actual chaotic oscillation in nonlinear regimes (see Figure 4.14).

Considering Equation 4.7, one can notice that the factor $(1+\tau/\theta)$ is easily replaced by 1 with the actual timescales. Another remark on this equation is concerned with the fact that a bandpass dynamics is practically removing any DC component after the transient. This leads to $\langle\varphi\rangle_t = 0$ and is forcing the constant of integration to be $I_0 = I_\beta \sin^2\varphi_0$ so that the term on the right-hand side of the equation also reveals a zero mean value. Due to the extremely spread characteristic timescales, one could think a priori that the integral term can be neglected: this is far from being correct, and special care has to be taken when simplifying Equation 4.7. Indeed, very slow transients or envelope dynamics of the order of the slowest response time θ are definitively observed in experiments, as we will see later.

In a first attempt to analyze the broadband bandpass dynamics and its first steady-state bifurcation, the corresponding dynamics can then be rewritten in a more conventional differential form (derivating the equality in order to remove the integration term), using thus a two-dimensional representation through the introduction of an intermediate dynamical variable u and renormalizing the time with respect to τ_D (since $2\tau_D$ periodic cycles are expected to be observed) through $s = t/\tau_D$:

$$u' = \varphi, \tag{4.9}$$

$$\frac{\tau}{\tau_D}\varphi' = -\varphi - \frac{\tau_D}{\theta}u + \beta\left[\sin^2(\varphi(s-1)+\varphi_0) - \sin^2\varphi_0\right], \quad (4.10)$$

where ( )′ stands for time derivation with respect to s. Looking for the fixed points gives naturally the trivial solution $u = x = 0$ (due to the DC removal action of the bandpass feedback). From this stable fixed point situation, an increase in the feedback gain leads to the expected "nearly" square wave solution (when φ_0 is adjusted to operate along the negative slope of the nonlinear function). This solution is found when setting the term on the left-hand side of Equation 4.10 to zero (supported by the fact that the factor of the derivative is $\tau/\tau_D \ll 1$). This gives two solutions in φ, the two values of the square wave plateaus. The solution in u is a triangle-like waveform, and the period is 2 (in units of τ_D), as expected.

Among the other observed regimes of this broadband bandpass delay dynamics, one finds an unexpected and new one compared to the situation met with the standard low pass feedback. Low-frequency cycles can indeed be observed, with typical timescales of the order of θ. This low-frequency cycle is obtained for values of φ_0 leading to oscillation along the positive slope of the nonlinear function (on the contrary to the $2\tau_D$ cycles usually observed along the negative slope of the nonlinearity). This slow cycle can be derived analytically from the differential model in Equation 4.7, when introducing a normalized slow time $s = t/\theta$, and neglecting both the fast timescale $\varepsilon = \tau/\theta \simeq 10^{-6}$, and the delay $\delta = \tau_D/\theta$. Derivating the model in Equation 4.7, we obtain

$$\varphi' = v \quad \text{and} \quad \frac{\tau}{\theta}v' = -v - \varphi + \beta\left[\sin^2(\varphi(s-\delta) + \varphi_0)\right]',$$

and with the asumptions $\frac{\tau}{\theta} \simeq 0$ and $\delta \ll s$: $\quad (4.11)$

$$\varphi' = v \quad \text{and} \quad v = -\frac{\varphi}{1 + \beta\sin[2(\varphi+\varphi_0)]}. \quad (4.12)$$

Equation 4.12 is actually a correct description of the slow cycle, except during the fast transition layer of the cycle (when $\beta\sin[2(\varphi+\varphi_0)]$ is close to -1). This solution illustrates one of the new bifurcation scenarios exhibited by the integro-differential delay dynamics, due to the slow timescale integral term.

Figure 4.15 summarizes the previously described dynamical regimes issued from the first bifurcation of the zero stable steady state in the integro-differential delay dynamics. The left plot corresponds to a bifurcation diagram with φ_0 as the bifurcation parameter along the horizontal axis. The vertical cut of the diagram (at constant φ_0) is grey scaled to render the statistics of the actually observed physical variable I_φ, and the dynamics are corresponding here to a fixed feedback gain ($\beta \simeq 1.3$). Dark grey indicates zero probability of the observed intensity. The general shape of the nonlinear function involved in the dynamics ($\sin^2$ shape) can be recognized in the bifurcation diagram, because for such a low feedback gain, most of the dynamics are corresponding to the stable steady states with $\varphi = 0$, and thus $I_\varphi = I_\beta \sin^2\varphi_0$. Only the φ_0 range leading to high enough slope of the nonlinear function (negative or positive) exhibits regime other than the stable steady-state

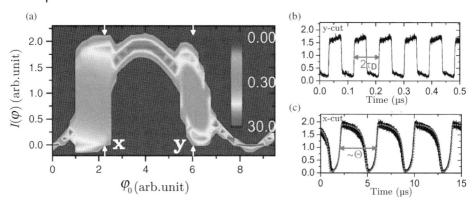

Figure 4.15 First bifurcations of the integro-differential delay dynamics (experiments, $\beta \simeq 1.3$). (a): bifurcation diagram; (b): c-cut of the left plot, $2\tau_D$ cycle (negative slope); (c): b-cut of the left plot (positive slope), low-frequency oscillation (gray dashed curve is numerics from Equation 4.12).

dynamics. The classical $2\tau_D$ cycle (middle plot in Figure 4.15) is obtained for a negative slope, whereas the low-frequency cycle (right plot) is attached to the positive slope of the nonlinear function. These regimes show the occurrence of very different timescales, indicating that each characteristic time in the equation can be relevant in the determination of the actual solution, depending on the parameter setting (β, φ_0).

When increasing further the bifurcation parameter, the different timescales start to interact, and this interaction leads to temporal waveforms combining at least two, or even all, of them together. Figure 4.16 shows typical experimentally observed examples of such a situation, with plots similar to those in Figure 4.15. The low-frequency oscillation occurred along the positive slope, showing within its envelope internal faster oscillations (lower middle plot). These oscillations are found to be of the order of $2\tau_D$ for the smallest oscillation amplitudes, or even faster where the amplitude is greater. The whole shape of the envelope together with this internal faster oscillation actually resembles to a bifurcation diagram of the corresponding low pass dynamics. It appears as if this bifurcation diagram is scanned at a slow timescale defined by θ. This is confirmed numerically by the dashed grey curve in Figure 4.16c.

Figure 4.17 illustrates the dynamical regimes obtained along the negative slope of the nonlinear function. Two typical regimes can be observed. One corresponds to fast oscillations with a pseudoperiod between τ and τ_D and is found for the upper part of the negative slope side of the nonlinear function (concave local shape of the nonlinear function). The other dynamical regime was already noticed earlier [19] for its strong multiple timescale complexity (see the middle waveform plots), and was referred to as "hyperchaotic breather." Qualitatively similar regimes can also be obtained from numerical simulations of the integro-differential model. The dynamical complexity within the slow envelope is illustrated in Figure 4.17d and 4.17f using wavelet transforms, thus illustrating the history of the waveform along the slow envelope of the dynamics.

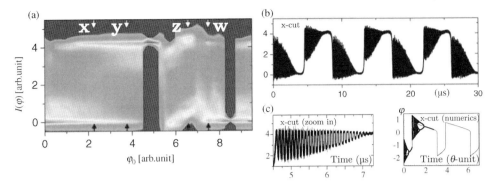

Figure 4.16 First mixed timescale dynamics (positive slope, $\beta \simeq 2.4$). (a): bifurcation diagram; (b,c): x-cut and its magnified waveform; (d): numerical simulation (plot of the extrema of the regime at $\beta = 2.5$) superimposed onto the low-frequency cycle for $\beta = 1.3$.

For high enough bifurcation parameters, the amplitude fluctuations are so strong (of the order of β) that no more local shape of the nonlinear function can be involved in the dynamics description (see Figure 4.18). The dynamics perform a global scanning of the nonlinear function, both on its positive and negative slope sides, flowing over one, two, or even three extrema of the $\sin^2$ nonlinearity. As it can be seen in Figure 4.18a, the statistics of the dynamics is only very weakly depending on φ_0. Fully developed chaos is thus obtained, which Fourier spectrum extends over the full available bandwidth of the broadband filter, and even more due to the nonlinear frequency spreading. It indicates a strong nonlinear mixing of the intrinsic timescales of the setup. This chaotic regime is a perfect one for the application of chaos communications that will be described later.

The next section will develop the second configuration of the bandpass electro-optic setup, where targeted application is requiring exactly the opposite signal properties compared to the chaotic one just reported: a pure sine wave oscillation, with an extremely high spectral purity.

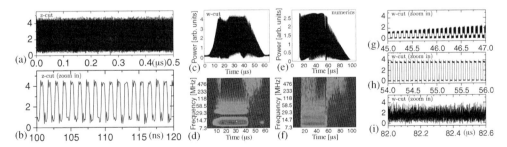

Figure 4.17 First mixed timescale dynamics (negative slope, $\beta \simeq 2.4$). (a,b): waveform at the z-cut in Figure 4.16 with two different timescales; (c–f): experimental (c) and numerical (e) full waveform corresponding to the w-cut in Figure 4.16, together with their wavelet transforms; (g–i): temporal details of the experimental dynamics at the w-cut in Figure 4.16, for different time shift within the envelope dynamics.

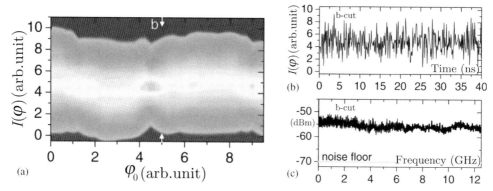

Figure 4.18 Strongly nonlinear regimes at β ≃ 4.3. (a): bifurcation diagram as in Figure 4.15; (b): temporal waveform obtained at the x-cut of the plot a); (c): corresponding RF spectrum.

Narrow Bandpass Filter: Huge Delay [41] In this configuration, the bandpass feedback filtering is resonant, thus strongly limiting the RF frequencies capable of oscillating within the delayed feedback loop. The sharpness of this resonant filter is described by a quality factor $Q = 1/(2\xi) = \Omega_0/(\Delta\Omega)$, with ξ being the damping factor, $F_0 = \Omega_0/(2\pi)$ the central microwave resonant frequency, and $\Delta F = \Delta\Omega/(2\pi)$ the $-3\,\mathrm{dB}$ bandwidth of the selective resonant filter. It corresponds typically to complex conjugate roots of the Fourier filter, leading to $\theta = \bar{\tau} = (\xi + i\sqrt{1-\xi^2})/\Omega_0$, which have the same modulus $1/(\Omega_0)$ corresponding to the central resonant frequency F_0 of the filter (Figure 4.14c), around 10 GHz. The position of the different characteristic frequencies attached to the narrowband feedback filter is illustrated in Figure 4.14c. It might be noticed that the characteristic time delay frequency $1/\tau_D$ stands out of the bandwidth of the filter, and is even extremely shifted in the low-frequency range due to the very long delay line used in that configuration (several kilometers).

One might think a priori that nonlinear phenomena will be strongly prevented due to the frequency-selective feature of the feedback (any harmonics generated by nonlinear effects will fall out of the filter bandwidth). The selectivity is however relative to the delay mode spacing, and the latter is indeed even much narrower than the bandwidth of the selective RF bandpass filter. A few kilometers of delay line lead to a few tens of kilohertz delay modes spacing, whereas the selective filtering is in the megahertz range. Many delay modes are thus potentially involved in the dynamics, and the nonlinear function of the feedback can still be strong, if the drive voltage allows a higher phase shift than π.

The model in Equation 4.7 is still valid, however with the previously described roots of the resonant filter τ and θ. Rewriting this dynamics and assuming that the solution will have an oscillating frequency around the center of the narrowband filter F_0, one can derive an envelope dynamics of the slowly varying amplitude $A(t)$ of the microwave oscillation $x(t) = A(t)\exp i\Omega_0 t$, ($x(t)$ being proportional to the electro-optically induced phase shift $\varphi(t)$):

$$\tau_A \frac{dA}{dt}(t) = -A(t) + \beta \sin(2\varphi_0) J_1(2A(t-\tau_D)). \tag{4.13}$$

The previous equation clearly reveals a low-pass delay dynamics on the envelope amplitude (on which nonlinear effects do occur), with a characteristic response time determined by the half-width of the narrowband RF filter, $\tau_A = 2/(\Delta\Omega)$. The long delay line authorizes many delay modes within the narrow bandwidth of the filter, meaning the envelope dynamics is still a large delay dynamics ($\tau_D \gg \tau_A$), for which a period doubling cascade is expected as the feedback gain is increased. This gain equals $\beta \sin(2\varphi_0)$, meaning it also depends on the operating condition of the Mach–Zehnder modulator (the offset phase of the two-wave interferometer). The maximum gain is achieved when the interference operates around the linear regime (where $\varphi_0 = \pm\pi/4$).

An important difference can be noticed when compared to the Ikeda low pass dynamics, which is concerned with the nonlinear transformation ruling the delayed feedback. Instead of the $\sin^2$ function, the Bessel function J_1 is obtained from the calculation. This Bessel nonlinearity is a consequence of the $\sin^2$ modulation (by the electro-optic interferometer) of the RF sine waveform at F_0. This leads to a Jacobi–Anger expansion, from which only the Bessel function J_1 is important, since it is the weight of the harmonics at F_0, the only relevant one in the feedback loop through the narrowband RF filter. The Bessel function J_1 however exhibits a similar nonlinear profile with respect to a $\sin^2$, at least for the first extremum.

The steady state of the envelope amplitude is the solution of the equation $A = \beta\sin(2\varphi_0)J_1(2A)$. Its stability can be analyzed the same way as the Ikeda low pass dynamics [41]. This gives a critical gain $\beta_{cr}\sin(2\varphi_0) \simeq 2.31$ (at the amplitude $A_{cr} \simeq 1.20$) at which a Hopf bifurcation occurs: below this gain the steady state is stable, and above it is unstable while a stable limit cycle is obtained. The limit cycle of the envelope corresponds for the RF oscillation to an amplitude modulation, with a modulation period of $2\tau_D$. The Hopf bifurcation of the envelope is thus a torus bifurcation of the RF oscillation.

Figure 4.19 shows the experimental bifurcation diagram for the envelope, obtained with an optoelectronic oscillator (OEO) at $F_0 = 3$ GHz with a 4 km delay line ($\tau_D = 20$ μs). Time traces and RF spectra before and after the critical gain are also represented. A theoretical diagram established according to the analytical results described below is represented in the same figure and shows a fairly good quantitative agreement with the experiment.

4.4 Applications

In this section, a few applications of optical delay dynamical systems will be briefly reported. Delay dynamics have the advantage to be very efficient in producing complexity, through strongly destabilized, but synchronizable, global behaviors, or on the contrary, they also can induce powerful stabilization in space or time.

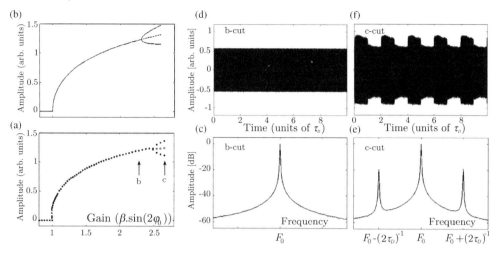

Figure 4.19 Bifurcation of the narrowband electro-optic delay dynamics. experimental (a) and theoretical (b) bifurcation diagram of the envelope; (c,d): temporal waveform and FFT spectrum before the critical gain; (c,f) temporal waveform and FFT spectrum after the critical gain.

4.4.1
Chaos Communications

Secure chaos communications was proposed as a straightforward application of the demonstrated possible synchronized behavior between two distant chaotic oscillations [45]. Synchronization is indeed a key feature in communication systems. Although the common synchronization occurs with sine waveforms, pseudorandom signals can a priori also be used as carriers for information encoding and transport. They require however more sophisticated techniques in order to achieve synchronization, as well as some "key" information, without which synchronization is difficult or possibly even impossible to obtain. Without this synchronization, the recovery of the information carried by the pseudorandom (e.g., chaotic) carrier is also assumed to be difficult or impossible. Digital communication has already applied these principles to digital secure communication, through the so-called code division multiple access (CDMA) communications (typically used in Global Positioning System, as well as in the third generation of mobile phone communication technology).

First attempts to demonstrate capability of chaos communications were performed with chaotic electronic circuits, performing low dimensional chaos [46, 47]. The development of chaos communication principles in optics allowed the use of the infinite dimensional phase space dynamics from all-optical or optoelectronic delay systems. After a few demonstrations and proof of principles [48–51], academic consortia tried to also demonstrate the potential of the principles for high bit rate secure optical communications. The OCCULT project successfully demonstrated Gb/s communications with 10^{-7} BER over more than 100 km, as well as operation capability of optical chaos communications on an installed fiber

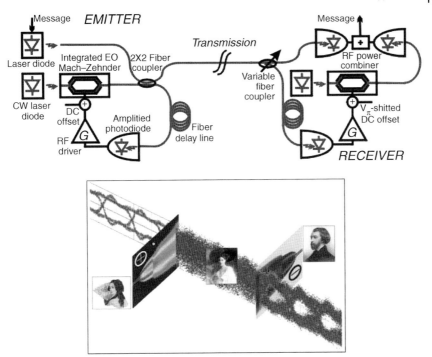

Figure 4.20 Intensity chaos for high bit rate optical communications. The chaotic waveform is generated by a broadband bandpass electro-optic delay dynamics.

optic network [43]. This was performed with both, an all-optical scheme and an electro-optical scheme.

For illustration, Figure 4.20 depicts the principle of the electro-optic emitter/receiver scheme that was used to demonstrate the 3 Gb/s intensity chaos communication system in Refs [42, 43]. The emitter is a closed loop, in which a binary intensity modulated signal is directly injected. The binary message thus contributes to the dynamics generated by the nonlinear delayed feedback loop. At the receiver side, the same nonlinear delayed dynamical processing is performed, however in an open loop configuration, sometimes also called self-synchronizing scheme. The message is recovered at the receiver side by subtracting the locally replicated dynamics from the directly received signal, the latter consisting of the superposition of the message and the chaos-like waveform generated by the nonlinear delayed feedback. The image on the right in Figure 4.20 is a visual interpretation of the chaos communication principle. The original clear message (represented as an open eye diagram on the left of the picture) is mixed in the chaos dynamics represented by a colorized bifurcation diagram. The resulting signal on the transmission channel is a completely closed eye (in the center of the picture), thus indicating that direct detection by an eavesdropper fails in recovering the clear message. At the receiver side, an inverse video colorized bifurcation diagram (in the right of the picture) represents the antiphase replication of the chaotic masking signal, which is required to recover the clear (but slightly blurred) original binary message. Added noise in the receiver open eye is due to the parameter mismatch between the emitter and receiver elements performing "nearly" similar nonlinear delayed dynamical processing.

Recently, an even more promising electro-optic phase chaos communication scheme [52] has been developed. It should offer a higher signal quality (error-free transmission) and at the same time a fastest speed capability (up to 10 Gb/s and higher).

Although chaos communication has demonstrated a clear applicable potential for physical layer secure high bit rate communications, the exact security level it can provide is still an open question. Future research projects related to chaos communications will definitely have to address the security issues. A combined physical algorithmic layer should allow the design of complex, but still physically synchronizable, chaos communication schemes, with a security supported by both the digital control of the chaos communication scheme and the chaotic motion itself.

4.4.2
Spatial or Spectral Stabilization in Semiconductor Lasers

4.4.2.1 High Spectral Purity Microwave Oscillation

It is surprising to realize that two applications of electro-optic delayed dynamics appeared nearly at the same time: chaos communication and high spectral purity microwave oscillations. It becomes even more surprising when noticing that both have nearly the same experimental architecture (Figure 4.14a), but strongly antagonistic signal properties (broadband pseudorandom white noise in one case and Dirac-like Fourier spectrum of a microwave sine waveform in the other case). The first application was described in Section 4.4.1 as an alternative way to secure optical information transmission at the physical layer. The second application is more concerned with time–frequency reference for radar, for which the Doppler shifted echo needs to be detectable in the sideband of the original microwave signal. To extract the small echo from the microwave noisy sidebands, the latter ones have to be as low as possible. From here comes the requirement for an ultrahigh spectral purity microwave signal in radar applications (short-term stability microwave reference). Quartz oscillators are well known for being excellent candidates in terms of high-stability time–frequency references (usually for long-term stability applications), but this stability is degraded at higher frequencies than 100 MHz, because of the necessity to multiply the intrinsic maximum quartz oscillators frequency from 100 MHz to the tens of gigahertz range. At the radar frequencies (typically 10 GHz and above), the electro-optic delay line oscillators are capable of superior performances, thanks to the technological progress done in the field of devices for high-speed optical communications: highly transparent propagation medium (glass fiber optics, about 0.2 dB/km, allowing ultralong delay lines compared to what is achievable with electrical coaxial cables), and extremely fast light modulators and light detectors (capable to operate up to tens of gigahertz frequencies).

From the physical principle point of view, the unusual feature of the optoelectronic oscillator consists in the storing energy element at the origin of the high spectral purity properties. This element is traditionally a resonator (quartz or dielectric), which exhibits a strong amplitude selectivity at the resonance, and at the same time a strong phase dependence with respect to frequency. The classical Barkhausen criteria for oscillators comprising such a resonator together with an amplifier to compensate for the losses of the passive resonator is usually met at the maximum of the resonance: a gain greater or equal to unity and an in-phase feedback for the actually oscillating frequency. The quality factor Q of the resonator alone is directly influencing the spectral purity of the oscillation. It is usually defined as $Q = F_0/\Delta F$ from the amplitude resonance curve of the resonator versus frequency (ΔF is the $-3\,\mathrm{dB}$ bandwidth from the maximum of the resonance). The quality factor is also related to the time domain, to the number of pseudoperiods observed in the damped oscillations of the resonator alone: the quality factor measures the capability of the resonator to store energy at the resonance frequency.

With a delay line OEO, the storing energy element is long fiber length of several kilometers, which is allowed by the very low absorption of modern optical fibers. These fibers are however not amplitude selective with respect to the microwave frequency carried by the traveling light beam: the extremely high equivalent quality factor is thus mainly related to the phase shift of the microwave frequency after having traveled through the delay line:

$$Q = \frac{F_0}{2} \cdot \frac{d\psi}{dF}. \tag{4.14}$$

With a fiber time delay τ_D, the microwave modulation of the optical carrier is phase shifted by a quantity $\psi = \Omega \tau_D = 2\pi F \tau_D$. This leads to an equivalent Q factor of $\pi F_0 \tau_D$, the expression that is very simple and yet illustrates very clearly why delay lines are preferably attractive for high-frequency oscillations, since the Q factor is linearly increasing with the microwave frequency. This expression of Q simply reveals that the time delay τ_D can "store" a number of microwave periods inversely proportional to the period or proportional to the frequency. It can be as high as 2×10^5 at $10\,\mathrm{GHz}$ for a $4\,\mathrm{km}$ delay line.

However, as already stated, the fiber delay line is not amplitude selective, so that a priori any delay mode can oscillate. A selective enough RF filter is thus required to roughly choose the actual oscillating frequency among all the possible delay modes. The longer the delay line is, the denser are the delay modes in the frequency domain (one mode every $50\,\mathrm{kHz}$ for a $4\,\mathrm{km}$ fiber delay line), thus justifying the role of the narrow RF filter in the OEO feedback loop. This leads to the same delay differential equation as the one in Equation 4.7. When noise performance of the OEO is concerned, this time domain delay differential equation is not only useful for stability analysis and in understanding the nonlinear effects but also for the definition of the spectral distribution of the actual noise in the OEO. This equation allows to derive a Langevin model for the noise influence on the complex envelope $\mathcal{A}(t) = |\mathcal{A}(t)|e^{i\psi(t)}$ of the microwave oscillation after a proper analysis of the actual noise sources and of their actions within the oscillation loop (multiplicative noise η_m and additive noise ζ_a (see Ref. [53]):

4 Optical Delay Dynamics and Its Applications

$$\frac{2}{\Delta\Omega}\frac{d\mathcal{A}}{dt}(t) = -\mathcal{A}(t) + 2[1+\eta_m(t)]\beta\sin(2\varphi_0)J_1(|\mathcal{A}(t-\tau_D)|)\,e^{i\psi(t-\tau_D)} + \zeta_a(t). \tag{4.15}$$

When the different noise diffusion constants and their spectral model are properly set with respect to the actual experimental situation, a theoretical phase noise spectral density $|\Psi(\omega)|^2$ of the microwave OEO signal can be derived from the Fourier transform phase of the dynamics $\psi(t) = \angle[\mathcal{A}(t)]$:

$$|\Psi(\omega)|^2 = \mu^2 \left| \frac{(2Q)^{-1}\tilde{\eta}_m(\omega) + |\mathcal{A}_0|^{-1}\tilde{\xi}_{a,\psi}(\omega)}{i\omega + \mu[1-e^{i\omega\tau_D}]} \right|^2, \tag{4.16}$$

where $\mathcal{A}_0$ is the steady-state solution of the microwave oscillation regime, at which we are looking for the phase noise features, $\tilde{\eta}_m(\omega) = \mathrm{FT}[\eta_a(t)]$, and $\tilde{\xi}_a(\omega) = \mathrm{FT}[\xi_a(t)]$.

The theoretical noise spectrum is plotted in Figure 4.21a, where the actual phase noise frequency range of the experimental result is indicated by the dashed red box. The theoretical spectrum can be compared with the measured one in Figure 4.21b. Both plots show very good agreement

- with a typical f^{-3} decay in the low frequency range,
- a noise floor together with the characteristic noise peaks of the nonoscillating delay modes at the multiple frequencies of $1/\tau_D$, and
- for frequencies out of the RF filter bandwidth ($f > \Delta\Omega/(4\pi)$), a f^{-2} decay.

The inset in Figure 4.21b is also a nice confirmation of the detailed sideband noise power spectrum: the theory, as illustrated in Figure 4.21a, predicts very sharp and high peaks for delay modes noise peaks, which cannot be resolved with the usual phase noise measurement resolution. When selecting properly the resolution around a small frequency window centered at the peak noise of the first nonoscillating delay mode, the measurement shows qualitatively, and even quantitatively, nearly the same values as the ones predicted by the phase noise model.

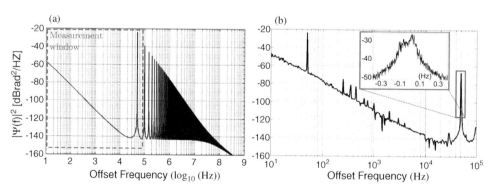

Figure 4.21 Phase noise spectra of an OEO. (a) Theoretical derivation with experiment-matched noise amplitudes. (b) Experimental result (*inset*: zoom of the first delay mode noise).

4.4.3
Further Applications of All-Optical Delay Systems

Coupled semiconductor lasers play an important role in many applications. Among them, longitudinal coupled-cavities lasers (e.g., C^3 lasers) have been used for spectral selection; coherent coupling allows high output power with good spectral and beam properties; laterally coupled laser arrays have also been realized to achieve coherent coupling. As presented in the previous sections, a delay in a feedback or coupling path introduces dynamical instabilities and particular synchronization properties that can be harnessed for applications. Here, we would only like to give a brief overview of the realized, suggested, or foreseen applications. Delayed feedback configurations have already been successfully utilized for chaos communication purposes, as discussed above. In addition, they have been utilized as light source for chaotic LIDAR [54] to generate fast random bit sequences [55] and to realize as light sources with over five orders of magnitude variable coherence length, which can be used for rainbow refractometry or optical coherence tomography [56]. The ambivalent character of delayed feedback, as discussed also for the electro-optic systems, is illustrated in applications like control of spatiotemporal instabilities of broad area semiconductor lasers [57]. But delayed-coupling configurations are also being considered for applications. In Ref. [58], a novel key exchange protocol has been suggested, utilizing the synchronization properties of two mutually delay-coupled semiconductor lasers with the semitransparent mirror as relay element. Finally, networks of delay-coupled lasers are currently being studied for the realization of novel information processing concepts, being inspired by neuronal systems. They are expected to represent a suitable reservoir for the realization of a liquid state machine [59]. All this demonstrates that delay effects that have been considered a nuisance for a long time are now being recognized as very useful, improving existing applications or allowing novel applications.

4.5
Conclusion and Outlook

Delay dynamical systems in optics show a rich phenomenology of dynamical properties and synchronization scenarios. They represent well-controllable test beds to study these systems that are being recognized to be of interest in more and more areas of science. Optical delay dynamical systems allow extreme features, depending on the chosen operating conditions: complex and broadband hyperchaos over a huge bandwidth or ultrastabilized (spatial or temporal) spectra. We illustrated the application potential of the reported configurations, as well as their tractable and scientifically rich theoretical description. The configurations presented in this study are only a few among many more. Motivated by the described coupling configurations, further and more complicated network arrangement of many delay-coupled optical oscillators represents a very promising and challenging field of study. Such studies might help us to understand some aspects of perception and information processing in the brain and to explore and realize the bio-inspired concepts of information processing like reservoir computing.

Acknowledgments

The authors would like to thank all those collaborators and institutions who contributed to this study, through their intellectual and financial help. They would like to express many thanks to Jean-Pierre Goedgebuer, Claudio Mirasso, Vladimir Udaltsov, Michael Peil, John Dudley, Wolfgang Elsäßer, Jan Danckaert, Raul Vicente, Miguel Cornelles Soriano, Guy Van der Sande, Thomas Erneux, Javier Martin-Buldu, Carme Torrent, Jordi Garcia-Ojalvo, Rajarshi Roy, Yanne Chembo, Pere Colet, Atsuchi Uchida, Maxime Jacquot, Jean-Marc Merolla, and Pierre Lacourt, as well as the different research programs from the European Community FP5 and FP6, the French ACI and ANR, CNET from France Telecom, the Institut Universitaire de France, the German Research Foundation, the Volkswagen Foundation, and the German BMBF.

References

1 Marchenko, Y. and Rubanik, V. (1965) Mutual synchronization of molecular oscillators. *Izvestiya VUZ. Radiofizika*, **8**, 679.

2 Marchenko, Y. (1967) Mutual synchronization of self-oscillating systems with consideration of the delay in interaction forces. *Izvestiya VUZ. Radiofizika*, **10**, 1533.

3 Kouda, A. and Mori, S. (1981) Analysis of a ring of mutually coupled van der Pol oscillators with coupling delay. *IEEE Trans. Circuits Syst.*, **CAS-28**, 247.

4 Orosz, G., Wilson, R.E., Szalai, R., and Stepan, G. (2009) Exciting traffic jams: nonlinear phenomena behind traffic jam formation on highways. *Phys. Rev. E*, **80**, 046205.

5 Yeung, M.K. and Strogatz, S.H. (1999) Time delay in the Kuramoto model of coupled oscillators. *Phys. Rev. Lett.*, **82**, 648.

6 Masoller, C., Marti, A., and Zanette, D. (2003) Different regimes of synchronization in nonidentical time-delayed maps. *Physica A*, **325**, 186191.

7 Erneux, T. (2009) *Applied Delay Differential Equations*, Springer, New York.

8 Balachandran, B., Kamr-Nagy, T., and Gilsinn, D. (2009) *Delay Differential Equations, Recent Advances and New Directions*, Springer, New York.

9 Michiels, W. and Niculescu, S.-I. (2007) *Stability and Stabilization of Time-Delay Systems. An Eigenvalue Based Approach, Advances in Design and Control*, SIAM, 12.

10 Stepan, G. (ed.) (2009) Theme issue: delay effects in brain dynamics. *Phil. Trans. R. Soc. A*, **367**, 1059.

11 Wolfram Just, M.S., Pelster, A., and Schöll, E.(eds) (2009) Theme issue delayed complex systems. *Phil. Trans. R. Soc. A*, **368**, 303.

12 Lang, R. and Kobayashi, K. (1980) External optical feedback effects on semiconductor injection laser properties. *IEEE J. Quantum Electron.*, **16**, 347.

13 Ikeda, K. (1979) Multiple-valued stationary state and its instability of the transmitted light by a ring cavity system. *Opt. Commun.*, **30**, 257.

14 Gibbs, H.M., Hopf, F.A., Kaplan, D.L., and Schoemaker, R.L. (1981) Observation of chaos in optical bistability. *Phys. Rev. Lett.*, **46**, 474.

15 Neyer, A. and Voges, E. (1982) Dynamics of electrooptic bistable devices with delayed feedback. *IEEE J. Quantum Electron.*, **18**, 2009.

16 Broom, R.F., Mohn, E., Risch, C., and Salathe, R. (1970) Microwave self-modulation of a diode laser coupled to an external cavity. *IEEE J. Quantum Electron.*, **6**, 328.

17 Neyer, A. and Voges, E. (1982) High-frequency electro-optic oscillator using an integrated interferometer. *Appl. Phys. Lett.*, **40**, 6.

18 Peil, M. *et al.* (2009) Routes to chaos and multiple time scale dynamics in broadband bandpass nonlinear delay electro-optic oscillators. *Phys. Rev. E*, **79**, 026208.

19 Kouomou, Y.C. *et al.* (2005) Hyperchaotic breathers in delayed optical systems. *Phys. Rev. Lett.*, **95**, 043903.

20 Larger, L., Goedgebuer, J.-P., and Merolla, J.-M. (1998) Chaotic oscillator in wavelength: a new set-up for investigating differential difference equations describing non linear dynamics. *IEEE J. Quantum Electron.*, **34**, 594.

21 van Tartwijk, G.H.M., Levine, A.M., and Lenstra, D. (1995) Sisyphus effect in semiconductor lasers with optical feedback. *IEEE Sel. Top. Quantum Electron.*, **1**, 466.

22 Levine, A.M., van Tartwijk, G.H.M., Lenstra, D., and Erneux, T. (1995) Diode lasers with optical feedback: stability of maximum gain mode. *Phys. Rev. A*, **52**, R3436.

23 Sano, T. (1994) Antimode dynamics and chaotic itinerancy in the coherence collapse of semiconductor lasers with optical feedback. *Phys. Rev. A*, **50**, 2719.

24 Fischer, I., van Tartwijk, G.H.M., Levine, A.M., Elsäßer, W., Göbel, E.O., and Lenstra, D. (1996) Fast pulsing and chaotic itinerancy with a drift in the coherence collapse of semiconductor lasers. *Phys. Rev. Lett.*, **76**, 220.

25 Heil, T., Fischer, I., Elsäßer, W., Mulet, J., and Mirasso, C.R. (2001) Chaos synchronization and spontaneous symmetry-breaking in symmetrically delay-coupled semiconductor lasers. *Phys. Rev. Lett.*, **86**, 795.

26 Fujino, H. and Ohtsubo, J. (2001) Synchronization of chaotic oscillations in mutually coupled semiconductor lasers. *Opt. Rev.*, **8**, 351.

27 Wünsche, H.-J., Bauer, S., Kreissl, J., Ushakov, O., Korneyev, N., Henneberger, F., Wille, E., Erzgrber, H., Peil, M., Elsäßer, W., and Fischer, I. (2005) Synchronization of delay-coupled oscillators: a study of semiconductor lasers. *Phys. Rev. Lett.*, **94**, 163901.

28 Mulet, J., Mirasso, C., Heil, T., and Fischer, I. (2004) Synchronization scenario of two distant mutually coupled semiconductor lasers. *J. Opt. B Quantum Semiclassical Opt.*, **6**, 97.

29 D'Huys, O., Vicente, R., Danckaert, J., and Fischer, I. (2010) Amplitude and phase effects on symmetry-breaking of delay-coupled oscillators. submitted to Chaos.

30 Roelfsema, P.R., Engel, A.K., König, P., and Singer, W. (1997) Visuomotor integration is associated with zero time-lag synchronization among cortical areas. *Nature*, **385**, 157.

31 Fischer, I., Vicente, R., Buldu, J.M., Peil, M., Mirasso, C.R., Torrent, M.C., and Garcia-Ojalvo, J. (2006) Zero-lag long-range synchronization via dynamical relaying. *Phys. Rev. Lett.*, **97**, 123902.

32 Vicente, R., Fischer, I., and Mirasso, C.R. (2008) Synchronization properties of three delay-coupled semiconductor lasers. *Phys. Rev. E*, **78**, 066202.

33 Peil, M., Larger, L., and Fischer, I. (2007) Versatile and robust chaos synchronization phenomena imposed by delayed shared feedback coupling. *Phys. Rev. E*, **76**, 045201(R).

34 Flunkert, V., D'Huys, O., Danckaert, J., Fischer, I., and Schöll, E. (2009) Bubbling in delay-coupled lasers. *Phys. Rev. E*, **79**, 065201(R).

35 Vicente, R., Gollo, L.L., Mirasso, C.R., Fischer, I., and Pipa, G. (2008) Dynamical relaying can yield zero time lag neuronal synchrony despite long conduction delays. *Proc. Natl. Acad. Sci. USA*, **105**, 17157.

36 Udaltsov, V.S. *et al.* (2002) Chaotic bandpass communication system. *IEEE Trans. Circuits Syst.*, **49**, 1006.

37 Mackey, M.C. and Glass, L. (1977) Oscillation and chaos in physiological control systems. *Science*, **197**, 287.

38 Le Berre, M. *et al.* (1987) Conjecture on the dimension of chaotic attractors of

delayed-feedback dynamical systems. *Phys. Rev. A*, **35**, 4020.

39 Erneux, T., Larger, L., Lee, M.W., and Goedgebuer, J.-P. (2004) Ikeda Hopf bifurcation revisited. *Physica D*, **194**, 49.

40 Larger, L. et al. (2005) From flow to map in experimental high dimensional electro-optic nonlinear delay oscillator. *Phys. Rev. Lett.*, **95**, 043903.

41 Kouomou, Y.C. et al. (2007) Dynamical instabilities of microwaves generated with optoelectronic oscillators. *Opt. Lett.*, **32**, 2572.

42 Larger, L. and Goedgebuer, J.-P. (2004) Encryption using chaotic dynamics for optical telecommunications. *C.R. De Physique*, **4** (5), 609.

43 Argyris, A. et al. (2005) Chaos-based communications at high bit rates using commercial fiber–optic links. *Nature*, **438** 343.

44 Yao, X.S. and Maleki, L. (1994) High frequency optical subcarrier generator. *Electron. Lett.*, **30**, 1525; Yao, X.S. and Maleki, L. (1996) Optoelectronic oscillator for photonic systems. *IEEE J. Quantum Electron.*, **32**, 1141.

45 Pecora, L.M. and Carroll, T.L. (1990) Synchronization in chaotic systems. *Phys. Rev. Lett.*, **64**, 821.

46 Kocarev, Lj. et al. (1992) Experimental demonstration of secure communications via chaotic synchronization. *Int. J. Bifurcat. Chaos*, **2**, 709.

47 Cuomo, K.M. and Oppenheim, A.V. (1993) Circuit implementation of synchronized chaos with applications to communications. *Phys. Rev. Lett.*, **71**, 65.

48 François, P.-L., Goedgebuer, J.-P., Larger, L., and Porte, H. (1990) French and European patent, extended to US, Système de transmission optique mettant en úuvre un cryptage par chaos déterministe FR 2743459, January 1996, France Telecom.

49 Van Wiggeren, G.D. and Roy, R. (1998) Communicating with chaotic lasers. *Science*, **279**, 1198.

50 Goedgebuer, J.-P., Larger, L., and Porte, H. (1998) Optical cryptosystem based on synchronization of hyperchaos generated by a delayed feedback tunable laser diode. *Phys. Rev. Lett.*, **80**, 2249.

51 Fischer, I., Liu, Y., and Davis, P. (2000) Synchronization of chaotic semiconductor laser dynamics on sub-ns timescales and its potential for chaos communication. *Phys. Rev. A*, **62**, 011801(R).

52 Lavrov, R. et al. (2009) Electro-optic delay oscillator with non-local non linearity: optical phase dynamics, chaos, and synchronization. *Phys. Rev. E*, **80**, 026207.

53 Chembo, Y.K. et al. (2009) Determination of phase noise spectra in optoelectronic microwave oscillators: a Langevin approach. *IEEE J. Quantum Electron.*, **45**, 178.

54 Lin, F.Y. and Liu, J.M. (2004) Chaotic lidar. *IEEE J. Sel. Top. Quantum Electron.*, **10**, 991.

55 Uchida, A., Amano, K., Inoue, M., Hirano, K., Naito, S., Someya, H., Oowada, I., Kurashige, T., Shiki, M., Yoshimori, S., Yoshimura, K., and Davis, P. (2008) Fast physical random bit generation with chaotic semiconductor lasers. *Nat. Photonics*, **2**, 728.

56 Peil, M., Fischer, I., Elsäßer, W., Bakic, S., Damaschke, N., Tropea, C., Stry, S., and Sacher, J. (2006) Rainbow refractometry with a tailored incoherent semiconductor laser source. *Appl. Phys. Lett.*, **89**, 091106.

57 Mandre, S.K., Fischer, I., and Elsäßer, W. (2003) Control of the spatiotemporal emission of a broad-area semiconductor laser by spatially filtered feedback. *Opt. Lett.*, **28**, 1135.

58 Vicente, R., Fischer, I., and Mirasso, C.R. (2007) Simultaneous bidirectional message transmission in a chaos-based communication scheme. *Opt. Lett.*, **32**, 403.

59 http://ifisc.uib-csic.es/phocus project webpage.

5
Symbolic Dynamics in Genetic Oscillation Patterns
Simone Pigolotti, Sandeep Krishna, and Mogens H. Jensen

5.1
Introduction

Symbolic dynamics has proven to be a very useful tool for studying many aspects of dynamical systems theory. The basic idea of symbolic dynamics is to study global properties of dynamical systems by dividing the phase space into sectors and analyzing the properties of the sequence of sectors visited by trajectories. From the first application by Hadamard to geodesic flows [1], this concept has found a number of fruitful applications [2], from the formal theory of dynamical systems to bridging dynamical systems and information theory [3].

More recently, this kind of approach has been applied to dynamical systems inspired by biology [4–7]. The description of biological systems as dynamical systems has proven to be very useful [8], but there are usually a number of problems related to our poor knowledge of the system parameters and the initial conditions. In this framework, symbolic methods can contribute to analysis of dynamical systems in a general way, without depending too much on the parameter details.

In this chapter, we will discuss a recent method [6, 7] that can be applied to dynamical systems with monotone interactions, in which variables activate or repress each other in a specified way. The method, which may be seen as generalization of the nullcline analysis, predicts that the order of maxima and minima of the variables has to comply with a number of constraints. We will show that this kind of approach can be used both to analyze dynamical systems and to face the more ambitious reverse engineering problem, that is, to figure out the best dynamical model from short experimental time series.

The chapter is organized as follows: Section 5.1 describes the general idea of the method. Section 5.2 goes more deeply into details of the particular case of a single negative feedback loop, where the symbolic dynamics is unique, and shows how the method can be used for reverse engineering. Section 5.3 discusses the more general case of multiple feedback loops, where chaotic dynamics may appear. Section 5.4 presents future perspectives on spatially extended regulatory systems. Section 5.5 is devoted to conclusions.

The Complexity of Dynamical Systems. Edited by J. Dubbeldam, K. Green, and D. Lenstra
Copyright © 2011 WILEY-VCH Verlag GmbH & Co. KGaA, Weinheim
ISBN: 978-3-527-40931-0

5.2
The Method

Generally, symbolic dynamics considers a dynamical system of the form

$$\dot{\mathbf{x}} = \mathbf{f}(\mathbf{x}),\tag{5.1}$$

where $\mathbf{x}, \mathbf{f}$ are vectors in $\Gamma \subset \Re^N$, and divides the phase space Γ into a number of sectors $\Gamma_1, \Gamma_2, \ldots, \Gamma_k$ such that $\Gamma_i \cap \Gamma_j = \emptyset$ for $i \neq j$ and $\cup_i \Gamma_i = \Gamma$. Then, symbols are associated with each sector to identify it, for example, a natural number s between 1 and k. A continuous time trajectory $\mathbf{x}(t)$ can then be arranged into a sequence of symbols $s_1, s_2, \ldots, s_T$ encoding the sequence of sectors visited by the trajectory.

The symbolic description is useful when it is possible to make predictions about the transitions from one sector to another. For example, in simple chaotic dynamical systems one can show that the symbolic dynamics for some partitions is Markovian or even Bernoulli, where probabilities are defined with respect to the natural measure. When this is possible, quantities such as the Kolmogorov–Sinai entropy can be calculated very easily [9].

For biological and chemical applications, it is interesting to consider cases in which the symbolic dynamics predicts that only a small fraction of the total number of possible transitions are allowed. This "no-go" class of results can be very useful for providing a global qualitative analysis of dynamical systems. Independent of the application, the specific choice of the partition into sectors is crucial, as symbolic dynamics on different partitions of phase space may have very different properties.

5.2.1
Monotonicity and Nullclines

When dealing with dynamical modeling of biological systems, a recurrent problem is that parameters characterizing the interactions and the dynamics of the different variables are poorly known. Sometimes, the only available information about the interactions is their sign, that is, whether the increase in a variable leads to an increase or decrease in the growth rate of another variable.

A useful property of many of these systems is that the sign of the interactions, as defined above, is independent of the value of the variables. In ecology, predators do not become prey when their abundance changes. In genetic systems, activators typically do not become repressors as their concentration varies. If we assume that the field $\mathbf{f}$ is smooth, this translates to the fact that we will consider systems in which the off-diagonal elements of the Jacobian matrix, $\partial f_i / \partial x_j$ have a fixed sign in phase space. We will use the language of genetic circuits and say that variable j *activates* variable i if $\partial f_i / \partial x_j > 0$ and that variable j *represses* variable i if $\partial f_i / \partial x_j < 0$ (clearly if there is no interaction from j to i, then $\partial f_i / \partial x_j = 0$). We will refer to this property as *monotonicity*; notice that this definition of monotonicity is different from the one usually adopted in dynamical systems theory [10]. Notice also that we do not make any assumptions about the diagonal terms of the Jacobian matrix; in particular, they can change sign, for example, in logistic growth models.

Given the assumption above, we partition the phase space into 2^N sectors characterized by the property that each component of the field f_i has a fixed sign. We assign to each sector a symbol such as $(+, -, \ldots, +)$, representing which component of the field is positive and which is negative. We say that two sectors are *adjacent* if their symbols differ by only one sign. In the following discussion, we will only consider transitions between adjacent sectors. This requires a variable to change from increasing to decreasing (or vice versa) and is equivalent to determining when the variable can have a maximum or minimum (transitions involving more sign changes are unrobust since they involve the occurrence of extremal values of two variables at exactly the same time).

We note that adjacent sectors are separated by a *nullcline*, that is, a manifold satisfying $f_i(\mathbf{x}) = 0$, where i is the variable undergoing the sign flip. For example, a minimum for the variable x_i corresponds to a crossing of the nullcline $f_i = 0$ from the region $f_i < 0$ to the region $f_i > 0$. This is possible only if, somewhere on the nullcline, the scalar product between the vector field $\vec{f}$ and the vector ∇f_i (which is normal to the nullcline $f_i = 0$) is positive:

$$\sum_{j \neq i} f_j(x_1, x_2, \ldots, x_N) \, \partial_{x_j} f_i(x_1, x_2, \ldots, x_N) > 0. \qquad (5.2)$$

The $i = j$ term is excluded since it is zero on the nullcline. By assumption, all the derivatives have fixed signs, and in any given sector the f_j's also have fixed signs (encoded in the associated symbol). If the symbol and derivative signs are such that each term is negative, then the sum cannot be positive. This implies the following rule:

A variable cannot have a minimum if all its repressors are increasing and all its activators are decreasing.

Similarly, for maxima:

A variable cannot have a maximum if all its repressors are decreasing and all its activators are increasing.

By using the above two rules, we can construct a transition diagram for a given biological system, with one node for each symbol and an arrow for each transition that does not violate the above rules. In the following section, we will discuss applications of the method, both as a tool for qualitative analysis of biological systems and as a tool for reverse engineering.

5.3
The Negative Feedback Loop

In this section, we discuss the allowed transitions for a single negative feedback loop (NFL). Single NFLs, such as the ones shown in Figure 5.1, are possibly the simplest networks that are capable of oscillatory dynamics.

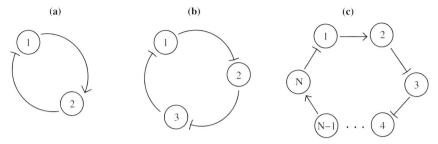

Figure 5.1 Examples of negative feedback loops: (a) the simplest case with two nodes, consisting of one activator and one repressor; (b) a three-repressor loop; (c) a general loop with N variables and an odd number of repressors.

We are then interested in understanding dynamical systems of the form

$$\dot{x}_i = f_i^{A,R}(x_i, x_{i-1}), \tag{5.3}$$

where the superscript in f_i denotes whether variable i is activated ($\partial f_i/\partial x_{i-1} > 0$) or repressed ($\partial f_i/\partial x_{i-1} < 0$), and the system is periodic, $x_0 = x_N$. To make things simpler, we here assume monotonicity also for the diagonal terms, that is,

$$\partial f_i/\partial x_i < 0 \quad \forall i. \tag{5.4}$$

This property is often satisfied by genetic systems (proteins are degraded at a faster rate at larger concentrations).

We start by searching for a fixed point $\mathbf{x}_*$ and studying its stability properties. The monotonicity allows us to invert the relations:

$$f_i(x_i^*, x_{i-1}^*) = 0 \rightarrow x_i^* = g(x_{i-1}^*). \tag{5.5}$$

By iterative substitution, we obtain

$$\begin{aligned} x_i^* &= g_i(x_{i-1}^*) = g_i(g_{i-1}(x_{i-2}^*)) = \cdots \\ &= g_i \circ g_{i-1} \circ g_{i-2} \circ \cdots \circ g_{i+1} \equiv F(x_i^*), \end{aligned} \tag{5.6}$$

where $\circ$ denotes convolution of functions. The function $F()$ is monotone and decreasing as a consequence of the derivative chain rule, $F'(x) = \prod_i g'(x_i^*)$ (remember that we have an odd number of decreasing f_i in NFL). As a consequence, Equation 5.6 cannot have more than one fixed point.

To perform the stability analysis, we write the characteristic polynomial evaluated at this point:

$$\prod_i [\lambda - \partial_x f_i(x, y)|_{x=x^*}] = \prod_i \partial_y f_i(x, y)|_{x=x^*}. \tag{5.7}$$

The above equation can be greatly simplified using the relation $F'(x) = \prod_i \partial_y f_i(x, y)/\partial_x f_i(x, y)$, which is a consequence of the implicit function theorem and the chain rule. One then obtains the following equation:

$$\prod_{i=1}^{N} \left(\frac{\lambda}{h_i}+1\right) = F'(x^*), \tag{5.8}$$

where the $h_i = -\partial_x f_i(x_i, x_{i-1})|_{x^*}$ denote the degradation rates at the fixed point. Notice that, because $F'(x)$ is always negative in a negative feedback loop, all coefficients of the characteristic polynomial are nonnegative; hence, it cannot have real positive roots. This means that the destabilization of the fixed point can occur only via a Hopf bifurcation, that is, with two complex conjugate eigenvalues crossing into the positive real half plane.

In the simple case in which all the degradation rates are equal and unchanging (i.e., $h_i = \gamma$, a constant), the roots of the polynomial (5.8) in the complex plane are the vertices of a polygon centered on $-\gamma$ with a radius $|F'|$ as sketched in Figure 5.2.

Therefore, the fixed point will remain stable as long as

$$|F'(x^*)|\cos(\pi/N) < \gamma. \tag{5.9}$$

In this case, Hopf's theorem (see, for example, [11]) ensures the existence of a periodic orbit close to the transition value, whose period is

$$T = 2\pi/\mathrm{Im}(\lambda), \tag{5.10}$$

which, in the simple case of equal degradation timescales, becomes $T = 2\pi/[|F'(x^*)| \cdot \sin(\pi/N)]$. Notice that the Hopf theorem does not ensure the

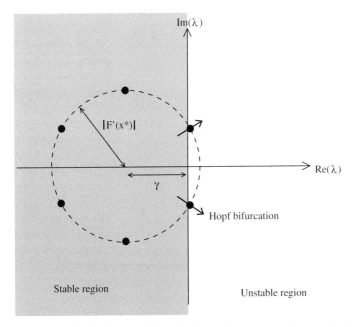

Figure 5.2 Sketch of the Hopf bifurcation in the eigenvalue complex plane, in the case in which all the degradation rates are equal to a constant γ.

orbit is stable; however, since the system is bounded and there are no other fixed points, we expect the orbit to be attracting, at least close to the transition point.

Our next goal is to describe how the space Γ is partitioned by the N nullclines μ_i defined by $f_i(x_i, x_{i-1}) = 0$. The properties we are about to state are all consequences of the monotonicity of the functions $f_i(x_i, x_{i-1}) = 0$, the constraint of having bounded and persistent orbits, and the existence of a unique fixed point $\mathbf{x}_*$.

It is worth remarking here that the existence of a fixed point is more a consequence of the boundedness condition than the monotonicity condition. Indeed, if the functions $g_i(x_{i-1})$ and $g_{i-1}^{-1}(x_{i-1})$ (see Equation 5.2) have independent support, they will obviously have no intersection. But in this case the system will not be persistent: persistence requires that $\lim_{x \to 0} f(x, y) > 0 \, \forall y$ and boundedness requires that $\lim_{x \to \infty} f(x, y) < 0 \, \forall y$; in this case, the nullclines have to cross. This fact is crucial also for the other considerations on the phase space portrait. Actually, the existence of a fixed point could have been demonstrated using only the boundedness and persistence hypothesis by means of Brouwer's fixed point theorem (see, for example, [12]).

Returning to the partitioning of Γ, the first important property is that any nullcline divides Γ into two simply connected sets, one in which $f_i(\mathbf{x}) > 0$ and one in which $f_i(\mathbf{x}) < 0$. Notice also that these manifolds cannot be tangent at the fixed point because of monotonicity and since they can depend at most on one common variable. All these properties imply that the Γ space is partitioned by the nullclines into 2^N simply connected sectors. In each of them, every component of the vector field has a definite, unchanging, sign. The uniqueness of the fixed point and the fact that the field has a constant sign inside a sector allows one to exclude the possibility of an attractor entirely contained within a sector. In the rest of this section, we discuss the case in which the fixed point is unstable. Since we assume that trajectories are bounded, starting from a sector the trajectory has to leave it by crossing one of the N boundaries.

We associate with a given symbol, or sector, the quantity H defined as the number of boundaries that can be crossed from that sector. Notice that $H = 0$ is impossible. For this to happen, the above rules must be violated by every variable, that is, the signs on both ends of an activation arrow must be the same, while the two signs on both ends of a repression arrow must be different. As there are an odd number of repressors, this is impossible. Therefore, $1 \leq H \leq N$. When the trajectory crosses a nullcline $f_i = 0$, as a simple consequence of the transition rules, H either stays constant, if the nullcline f_{i+1} can be crossed from the new sector, or decreases by two if f_{i+1} cannot be crossed (see Figure 5.3). Physically, if we think of a crossable boundary as "mismatch," that is, an unsatisfied bond between sign i and $i-1$, H is the number of such mismatches; hence, it quantifies the level of "frustration" in the system. The time evolution can then either (i) solve two neighboring unsatisfied bonds reducing H by two or (ii) shift an unsatisfied bond one place to the right, that is, from i to $i+1$, without changing H. As a consequence, H can never increase, and it must always be an odd number (see Figure 5.3). The conclusion is that we expect the system to end up in a state in which $H = 1$ and the unsatisfied bond keeps on moving around the loop in the direction of the arrows. This represents, at the level of symbolic

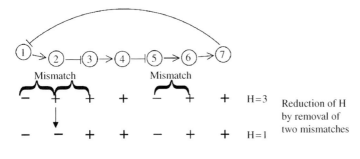

Figure 5.3 Consequences of crossing a nullcline for the sign vector and the quantity H. *Upper panel*: A sign is changed between two mismatches (in gray) and, consequently, H decreases by 2. *Lower panel*: A sign is changed next to just one mismatch. In this case, the mismatch simply moves one step ahead in the loop; the value of H does not change.

dynamics, a single "signal" traveling around the loop. A direct implication is that the extremal points (maxima or minima) of all variables should appear in the time series in the order in which the species are arranged in the cycle.

This is the simplest scenario and is the only possible one for $N < 4$. For $N \geq 4$, one cannot exclude that $H = 3$ can become a stable state, with three different mismatches traveling along the feedback loop. Notice that even if there is an attractor with $H = 3$, it must anyway coexist with an $H = 1$ attractor since if a trajectory ever starts from, or enters, a sector with $H = 1$, it can never return to an $H = 3$ sector. Furthermore, the $H = 3$ attractor is likely to require some fine-tuning of parameters to avoid one mismatch traveling "faster," catching up, and annihilating another one. It is thus likely that any perturbation or noise would bring the system to the "ground-state" attractor characterized by $H = 1$, which is the one we expect to observe in biological systems.

5.3.1
NFL and Reverse Engineering

We have shown how the single NFL generates a unique, robust symbolic dynamics. A very important property is that different NFLs generate different symbolic dynamics. This implies that each periodic symbolic sequence, if it is compatible with a NFL, allows us to identify the interactions in a unique way.

For example, we consider the p53-Mdmd2 oscillations observed in single-cell fluorescence experiments, shown in Figure 5.4a. We use the method to answer the

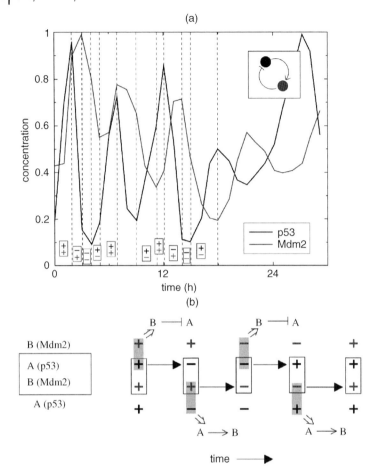

Figure 5.4 (a) p53-Mdm2 oscillations as recorded in a fluorescence microscopy experiment [13] and the reconstructed symbolic dynamics. *Inset*: The negative feedback loop extracted using the algorithm in the text; the process is shown in (b). Here, and in subsequent figures, ordinary arrows represent activation, while barred arrows represent repression.

question: is the dynamics compatible with the one generated by a single NFL? If so, what are the interactions?

The application of the method, based on the results of the previous subsection, can be reduced to the following algorithm:

1) List the order in which the maxima and minima of the variables occur. For example, in Figure 5.4a, the order is p53 max, Mdm2 max, p53 min, Mdmd2 min, p53 max, Mdm2 max, and so on.
2) If the variables occur in this list in an unchanging cyclic order, then this fixes the order of species in the loop, that is, a variable activates or represses the one

immediately following it in the list. Otherwise, a single negative feedback loop is inconsistent with the time series.

3) Construct the symbolic dynamics for the time series, with $+/-$ symbols listed in the order obtained in step 2.
4) If the symbolic dynamics is not periodic, a single negative feedback loop is inconsistent with the time series. Otherwise, start with any variable and the one pointing to it (say, variables B and A) and note the steps where B changes sign. If at the previous step (before B changed) A had the same sign as B, then A represses B, else it activates B (see Figure 5.4b).
5) This procedure is repeated for each variable to obtain the effect of the preceding variable. For example, in Figure 5.4, we conclude that p53 activates Mdm2 and Mdm2 represses p53. If the various sign changes in any variable give inconsistent conclusions, then a single negative feedback loop is inconsistent with the time series.
6) If the number of repressors in the loop is even, then a single negative feedback loop is inconsistent with the time series.

Therefore, we conclude (i) the observed oscillations are consistent with a monotone NFL and (ii) however, for this there must be at least one other unobserved species taking part in the loop since the fixed point is always stable for $N = 2$. Indeed, several three variable models of p53-Mdmd2 oscillations have been examined, which assume the third variable to be either an Mdm2 precursor (e.g., Mdm2 mRNA) or a third protein that interacts with p53 or Mdm2 [13]. Of course, it is hard to say at this level if a simple, deterministic negative feedback loop is a good model for this specific system. For example, the same sequences of symbols could be observed in a time delay model [14]. In addition, experiments suggest the presence of at least 10 feedback loops involving p53 [15]. It is also nontrivial to assess whether the irregular behavior of the trajectory is either due to internal mechanisms (chaos, time delays, etc.) or due to interaction with other proteins. Still, we can conclude that a single NFL is a good candidate for a zero-order model, being simple and reproducing correctly the qualitative behavior of the components with the correct interaction signs.

The next example involves circadian oscillations of gene expression in cyanobacteria. Cyanobacteria are the only bacterial species with a circadian clock and several of their cellular functions appear to be under circadian control [16]. In *Synechococcus elongatus*, a cluster of three genes, *kaiA,B,C*, were found to be essential: deletion of any of these genes eliminated the oscillations [17]. Figure 5.5a shows circadian rhythms in the expression levels of genes coding for homologues of the KaiA,B,C proteins in one *Synechocystis* strain. The symbolic dynamics is consistent with a three-variable feedback loop (Figure 5.5a, inset), where *kaiA* activates *kaiC1*, which represses *kaiB3*, which, in turn, activates *kaiA*. The first two of these predicted interactions exist in *Synechococcus* [17], while the third is a new prediction for how *kaiA* is brought into the loop. Note that our analysis provides only the sign of this interaction. It does not reveal the molecular mechanism of the interaction, or whether the interaction is direct or through intermediate steps.

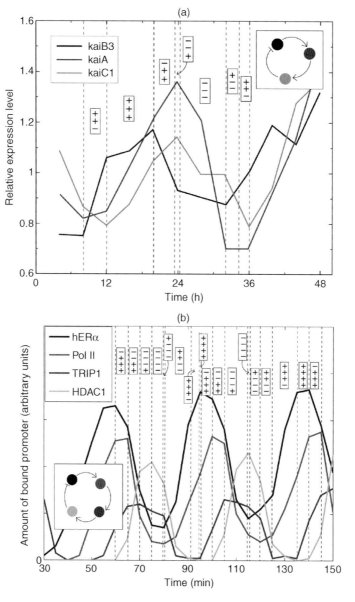

Figure 5.5 (a) Circadian rhythms of three *kai* genes in a *Synechocystis* cyanobacterial strain (data from Ref. [18]). (b) Periodic binding of four proteins to the pS2 promoter following addition of estradiol (data from Ref. [19], based on Ref. [20]). In each case, the corresponding symbolic dynamics is also shown, with symbols in the same order as the legend (where the maxima/minima of two variables occur very close; we have exaggerated the separation between the dotted lines for visual clarity). The insets show the topology deduced from the symbolic dynamics.

Finally, we consider the cyclic binding of cofactors to the estrogen-sensitive pS2 promoter. A coordinated sequence of binding and unbinding events modifies the DNA packing and nucleosome structure to enable transcription to proceed [20]. This is a case where no model exists and not all the proteins involved have been identified. Our method is particularly suited for such a case because it does not matter if the dynamics of only a subset of the proteins involved is available. Ref. [20] measured the temporal dynamics of binding of several proteins at the pS2 promoter using ChIP assays. Figure 5.5b shows oscillations in the binding of four proteins, after the addition of estradiol. The ER, estradiol receptor, binds estradiol and is required for initiating transcription. Pol II is the RNA polymerase that transcribes the gene. TRIP1 is a component of the APIS proteasome subunit, while HDAC1 is involved in deacetylation of histones [20]. The symbolic dynamics is consistent with the model shown in Figure 5.5b (inset). Notice that in this case, each variable measures the amount of bound protein at the pS2 promoter (albeit in arbitrary units, which are different for each protein). The predicted links indicate how a bound protein affects the probability of binding (or of remaining bound) of another one in the sequence. For example, the link from ER to Pol II indicates that ER, when bound at the promoter, increases the recruitment probability of Pol II. Ref. [19] models the dynamics of Figure 5.5b using a *positive* feedback loop, requiring over 200 intermediate steps, which has only activating links. Our analysis predicts the existence of a repressive link between HDAC1 and ER. One could imagine this repression as a result of a change in the conformational state of the DNA or histones, or the blocking of a binding site for ER at the promoter. However, we emphasize again that our analysis does not give any information about the molecular mechanism for this repression or whether there are intermediate steps. The analysis does suggest that for this system, a negative feedback loop is a plausible hypothesis as the cause of oscillations.

5.4
Multiple Loops

In this section, we discuss examples of circuits with more feedback loops. In these cases, the symbolic dynamics is not unique and one can expect a richer dynamical behavior, including chaotic dynamics.

We first consider the example network of Figure 5.6: a three-species negative feedback loop with a cross-link from node 3 to 2 that introduces a positive feedback. By checking all the allowed transitions, we construct the corresponding transition network, shown in Figure 5.1a (right). For example, from the symbol $(-++)$ the transition $(-++) \to (-+-)$ is ruled out because all the activators of node 3 are increasing; therefore, it cannot have a maximum. Similarly, we rule out all transitions from it except $(-++) \to (--+)$. The result, in this case, is a simple modification of the transition network for a single negative feedback loop shown by the solid arrows in Figure 5.6 [6]. With the cross-link present, the additionally allowed transitions are the ones shown with dashed arrows. The following dynamical system

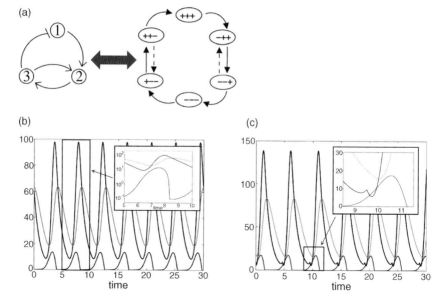

Figure 5.6 Simple example of the use of symbolic dynamics. (a) *Left*: Scheme of the network. *Right*: Corresponding transition network; we removed for clarity the symbols $(+-+)$ and $(-+-)$, which have no incoming links. (b) Dynamics of the three variables as a function of time with a weak cross-link ($a = 10$, see text), showing the transition cycle in solid arrows in (a). Inset shows the same on a log scale. (c) Dynamics for a stronger cross-link ($a = 50$, see text) that includes the transitions shown by dashed arrows, zoomed in the inset. In all plots x_1 is dark gray, x_2 is light gray, and x_3 is black.

illustrates these possibilities: $\dot{x}_1 = c - x_1 x_3/(k_1 + x_1)$; $\dot{x}_2 = x_1^2 + a[\theta(x_3 - k_2) - 1] - x_2$; $\dot{x}_3 = x_2 - x_3$. The major nonlinearity is the Heaviside step function: $\theta(x) = 0$ for $x < 0$ and $\theta(x) = 1$ for $x > 0$[1]. By choosing parameters such that the cross-link is weak ($c = 30$, $a = 10$, $k_1 = 0.1$, $k_2 = 20$), one obtains dynamics of Figure 5.1b, which is identical to the simple three-species loop. As the strength of the cross-link is increased ($a = 50$), the symbolic dynamics changes to also exhibit the dashed transitions. This is shown in Figure 5.6 where variable x_2 develops a new small maximum, thus changing the symbolic dynamics.

We now move on to a system that exhibits a richer range of dynamical behaviors (see Figure 5.7a). It consists of two negative feedback loops, coupled via a shared species. This network has been widely studied in the ecological literature [21–23] as a model for three trophic-level ecosystems: species x_3 feeds on x_2 and x_2 feeds on x_1. The chaotic properties of this motif have been used to interpret data from the Canadian lynx-hare cycle, showing irregular oscillations [24].

We consider first the Hastings–Powell (HP) model [22] as a dynamical system corresponding to this network:

---

1) We can safely introduce a discontinuous field because our argument works as long as trajectories are continuous.

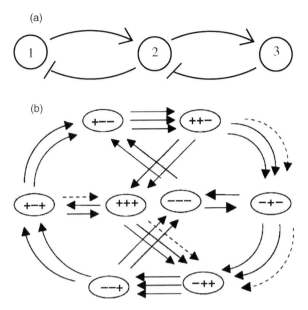

Figure 5.7 Network of two coupled two-species oscillators. (a) Structure of the network. (b) The transition network for this three-node system. Black arrows indicate all the allowed transitions. Dark gray arrows are the transitions actually observed in the HP system and light gray arrows are the transitions observed in the BHS model (see text). Dashed arrows indicate "kicks," that is, transitions *not* observed close to the Hopf bifurcation.

$$\begin{aligned} \dot{x}_1 &= rx_1(1-kx_1) - \alpha_1 x_1 x_2/(1+b_1 x_1) \\ \dot{x}_2 &= -d_1 x_2 + \alpha_1 x_1 x_2/(1+b_1 x_1) - \alpha_2 x_2 x_3/(1+b_2 x_2) \\ \dot{x}_3 &= -d_2 x_3 + \alpha_2 x_2 x_3/(1+b_2 x_2) \end{aligned} \quad (5.11)$$

with the following parameter choices: $\alpha_1 = \alpha_2 = 4$, $b_1 = b_2 = 3$, $d_1 = .4$, $d_2 = .6$, and $k = 1.5$. By increasing the parameter r, a stable limit cycle undergoes a series of period doubling bifurcations, followed by the onset of chaos. A projection of the attractor on the x_2–x_3 plane is shown in Figure 5.8. The chaotic trajectory looks similar to the periodic one, except for the irregular behavior of the amplitude [23]. This means that the same sequence as corresponding to the periodic orbit is observed after the onset of chaos. By increasing r even more, we found a regular window with a change in the symbolic dynamics (the "kick", shown in dark gray in the attractor in Figure 5.8 and in the bifurcation diagram, Figure 5.9a, and corresponding to the gray transition in Figure 5.2b). The kick is still present when, by further increasing r, the dynamics becomes chaotic again.

The conclusion is that the same symbolic dynamics observed close to the Hopf bifurcation is found in a large region of parameter space. We compare the results with a different system corresponding to the same network, the model by Blausius, Huppert, and Stone (BHS) [21]:

112 | 5 Symbolic Dynamics in Genetic Oscillation Patterns

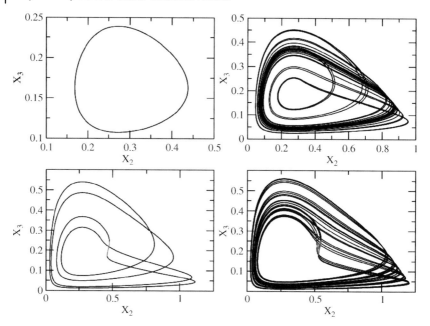

Figure 5.8 Two-dimensional projection of the attractor of the system of Equation 5.11 for different values of the control parameter $r = 2.0$ (*top left*), $r = 2.6$ (*top right*), $r = 3.0$ (*bottom left*), and $r = 3.3$ (*bottom right*).

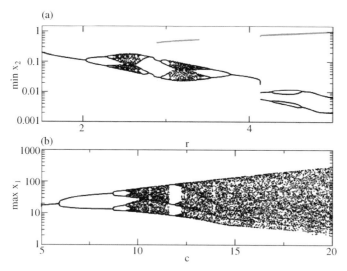

Figure 5.9 Bifurcation diagrams. *Top*: HP model (11), minima of x_2 plotted against r. *Bottom*: BHS model (12), maxima of x_1 versus c. In both plots, light gray indicates the appearance of "kicks" in the trajectory and symbolic dynamics (see text).

5.4 Multiple Loops

$$\dot{x}_1 = x_1 - \alpha_1 x_1 x_2/(1+kx_1)$$
$$\dot{x}_2 = -dx_2 + \alpha_1 x_1 x_2/(1+kx_1) - \alpha_2 x_2 x_3 \qquad (5.12)$$
$$\dot{x}_3 = c(x_3^* - x_3) + \alpha_2 x_2 x_3$$

with parameters $\alpha_1 = 2$, $\alpha_2 = d = 1$, $k = 0.12$, and $x_3^* = 0.006$. Here, a convenient control parameter is c. We observe the same scenario in the bifurcation diagram (see Figure 5.9b): periodic orbit, then chaotic but same periodic symbolic dynamics, then different symbolic dynamics in a regular window, and, finally, chaotic symbolic dynamics. Note, however, that the periodic symbolic dynamics observed close to the Hopf bifurcation is *different* from that observed in the HP model.

To test the robustness of the two sequences, we tried to change the functional form of the interaction between x_2 and x_3 by setting $b_2 = 0$ in the HP model or, conversely, introducing saturated response in the BHS model. We also tried to vary the parameters of both systems, by up to 50% from their default values. The two symbolic sequences were not affected by any of these changes. A possible cause for this robust difference could be the logistic term in the first part of (5.11), acting as a regulator so that the full dynamics is bottom-up controlled.

5.4.1
Multiple Loops and Reverse Engineering

The difference between the symbolic dynamics of the HP and BHS systems can be used for model selection: in the example of the Canadian lynx system, although one has access only to the lynx population time series, temporal measurements of the hare and grass abundances could be used to understand which model is more appropriate. Interestingly, from the point of view of maxima/minima order, these two systems behave like two different, single negative feedback loops [6]: 3⊣2⊣1⊣3 (HP system) and 1 → 2 → 3⊣1 (BHS system). Both these "effective" loops would include an "effective" interaction between variables x_1 and x_3.

To further elaborate on the reverse engineering problem, let us consider again the circadian oscillations of the three genes *kai A, B, C* in cyanobacteria discussed in Section 5.3.1, which had the following symbolic dynamics (B, A, C): $(+++) \to (-++) \to (--+) \to (---) \to (+--) \to (++-) \to (+++)$. Several networks are consistent with this pattern – the simplest is the loop $B \to A \to C \dashv B$, as suggested in Section 5.3.1. However, other more complicated networks can also give the same pattern and thus there is no unique way available to reverse engineer the underlying network. However, by combining the information from our method with other experimentally known facts, we can nevertheless obtain some nontrivial guidelines for which further interactions to experimentally look for. Specifically, experiments have shown that $A \to C$ and $C \dashv B$ and that all three genes are essential for oscillations [17]. Adding this information to the constraints obtained from our method leads us to conclude that (i) *kaiA* must be either activated by *kaiB* or repressed by *kaiC* (or both) and (ii) if *kaiA* is not activated by *kaiB*, then, in addition to *kaiC*⊣*kaiA*, *kaiB* must activate *kaiC*, so that the underlying network looks similar to Figure 5.6a. Of course, these predictions are for "effective" interactions, which at the molecular level could involve multiple intermediates, such as chemical complexes and various protein activity states. Ref. [6] shows how the method can reconstruct effective interactions even in the presence of intermediate species.

This circadian example also points out how much information our method provides. The transition $(+\,+\,+) \rightarrow (-\,+\,+)$ means that either $B \dashv A$ or $C \dashv A$. Later transitions show that either $A \rightarrow B$ or $C \dashv B$ and $A \rightarrow C$ or $B \rightarrow C$. Even without the extra experimental information, our method reduces the possibilities for the adjacency matrix of the underlying network from 3^6 to 5^3, a factor of ≈ 6. In a general N node system, with M independent observed transitions, the fraction of allowed adjacency matrices is $[1-(2/3)^{N-1}]^M$; the smaller the network and the more the transitions seen, the more useful the method. A full oscillation cycle would show at least N independent transitions. If the system instead reaches a fixed point, the transient can still be used.

5.5
Negative Feedback Loops on a Lattice

We have so far been concentrating on single feedback loops to model, for instance, what happens inside single cells. This means that the variables are only functions of time and not of space. In order to consider interactions between feedback loops, to model, for instance, cell–cell interactions [25], we present an idealized case where feedback loops interact on a lattice. Specifically, we choose repressilators [26], each consisting of three repressing genes, and place them on a hexagonal lattice. This is illustrated in Figure 5.10 where many repressilators are placed in such a way that they share repressor links between different nodes [27]. As seen in the figure, it is possible to construct such a repressor lattice without internal frustration in the sense that each repressilator motif acts in the usual way with three links between three nodes. The coupling between each motif, however, gives rise to very interesting dynamics. This

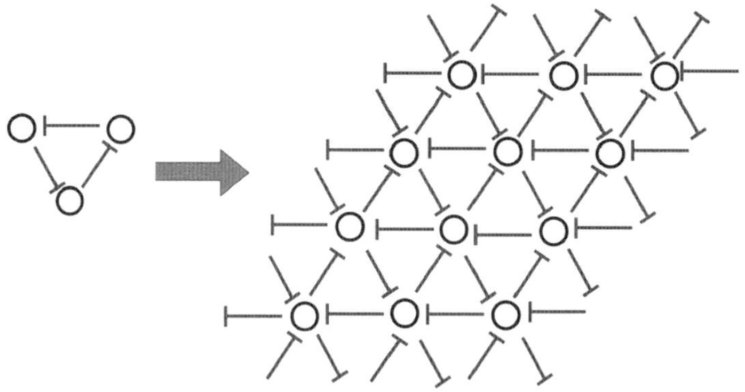

Figure 5.10 The construction of the repressor lattice from "units" of single repressilators suitably placed on a hexagonal lattice. Each link symbolizes a repressor between two nodes corresponding to repressing genes, proteins, species, and so on.

type of idealized lattice of interacting cells might be realized in biological tissues, such as onionskin or the retina, where it is well known that cells are organized in a hexagonal structure.

The dynamical equations for this repressor lattice are formulated in terms of the local variable at node (m, n), which is repressed by three neighboring nodes represented by an interaction term F_{int} [27]

$$\frac{dx_{m,n}}{dt} = c - \gamma x_{m,n} + \alpha F_{\text{int}}, \qquad (5.13)$$

where the interaction term is

$$F_{\text{int}} = \frac{1}{1 + (x_{m+1,y}/K)^h} \cdot \frac{1}{1 + (x_{m,n-1}/K)^h} \cdot \frac{1}{1 + (x_{m-1,n+1}/K)^h}. \qquad (5.14)$$

As usual for a lattice of interacting variables, the boundary conditions are important. For the results discussed here, periodic boundary conditions have been applied. Figure 5.11 shows results from a simulation of Equation 5.14 for a lattice consisting of 3×3 nodes. We observe periodic oscillations with three different solutions each differing by a phase equal to $2\pi/3$. This means that the dynamical solution at any of the nine nodes in the lattice belong to one of the solutions shown in Figure 5.11 and each repressilator therefore behaves in the "usual way" and there is no internal frustration in the lattice. This changes as the lattice becomes larger. Due to commensurability effects, symmetries are broken and more complicated solutions appear [27]. In a 5×5 lattice, just above the Hopf bifurcation where oscillations set in, there are in fact solutions with five distinct phases. Furthermore, there exist several such solutions due to the fact that the rotational symmetry is broken. As the interaction strength between the nodes in the lattice increases, the simple five-phase oscillations disappear, while the dynamical solutions become more and more distorted and finally end up in chaotic dynamics. All these transitions are indicated in the bifurcation diagram in Figure 5.12, showing standard periodic doubling transitions eventually ending up in chaotic dynamics.

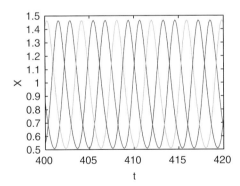

Figure 5.11 Solution from a 3×3 repressor lattice with multiplicative interactions, Equation 5.14 with parameters $c = 0.1$, $\gamma = 1.0$, $\alpha = 3.0$, $K = 1.0$, $h = 2$. Clearly, three different phases exist.

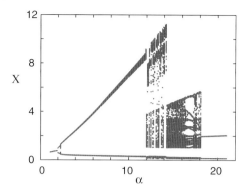

Figure 5.12 Bifurcation diagram for a 5×5 lattice with $c = 0.1$, $\gamma = 1.0$, $K = 1.0$, $h = 3$, and the multiplicative coupling. Shown are the maxima and minima for dynamical solutions in the node point $(5, 4)$ after a transient period of 15 000 time units. The extremal values are plotted against varying values of the coupling strength α.

5.6
Conclusions

We presented a method to construct a symbolic transition network that imposes a strong constraint on the dynamics of monotone systems, like many biological modules. In the case of a single negative feedback loop, we have shown how the method predicts that the symbolic dynamics is unique, making it very useful for reverse engineering. When more loops are present, uniqueness is lost and network reconstruction may be achieved only by complementing the method with other information, such as an experimental knowledge of some of the interactions. In all the cases we studied, the periodic symbolic dynamics observed close to the Hopf bifurcation is found in a large region of parameter space, even when the system becomes chaotic. This explains the commonly observed phenomenology of a chaotic attractor consisting of oscillations with randomly varying amplitude [23]. The oscillatory systems we looked at produce a symbolic sequence identical to that of a single negative feedback loop for most studied parameter values. By identifying these loops, our method can be used to derive minimal models of complex, oscillatory biological systems.

References

1 Hadamard, J. (1898) Les surfaces a courbures opposees et leurs lignes geodesiques. *J. Math. Pures Appl.*, **4**, 27–73. Reprinted in *Oeuvres de Jacques Hadamard*, vol. 2, CNRS, Paris, France, 1978, pp. 729–775.

2 Lind, D. and Marcus, B. (1995) *An Introduction to Symbolic Dynamics and Coding*, Cambridge University Press.

3 Shannon, C. (1948) A mathematical theory of communication. *Bell Syst. Tech. J.*, **27**, 379–423.

4. Glass, L. and Kaufmann, S.A. (1973) The logical analysis of continuous, non-linear biochemical control networks. *J. Theo. Bio.*, **39** (1), 103–129.
5. Glass, L. (1975) Combinatorial and topological methods in nonlinear chemical kinetics. *J. Chem. Phys.*, **63** (4), 1325–1335.
6. Pigolotti, S., Krishna, S., and Jensen, M.H. (2007) Oscillation patterns in negative feedback loops. *Proc. Natl. Acad. Sci.*, **10416**, 6533–6537.
7. Pigolotti, S., Krishna, S., and Jensen, M.H. (2009) Symbolic dynamics in biological feedback networks. *Phys. Rev. Lett.*, **102**, 088701.
8. Tiana, G., Krishna, S., Pigolotti, S., Jensen, M.H., and Sneppen, K. (2007) Oscillations and temporal signalling in cells. *Phys. Biol.*, **4**, R1–R17.
9. Beck, C. and Schlogl, F. (1993) *Thermodynamics of Chaotic Systems: An Introduction*, Cambridge University Press.
10. Smith, H.L. (1995) *Monotone Dynamical Systems: An Introduction to the Theory of Competitive and Cooperative Systems*, American Mathematical Society.
11. Murray, J.D. (2004) *Mathematical Biology*, Springer.
12. Hofbauer, J. and Sigmund, K. (1998) *Evolutionary Games and Population Dynamics*, Cambridge University Press.
13. Geva-Zatorsky, N. *et al.* (2006) Oscillations and variability in the p53 system. *Mol. Sys. Biol.*, **2**. doi: 10.1038/msb4100068.
14. Tiana, G., Jensen, M.H., and Sneppen, K. (2004) Time delay as a key to apoptosis induction in the p53 network. *Eur. J. Phys. B*, **29** (1), 135–140.
15. Harris, S.L. and Levine, H.J. (2005) The p53 pathway: positive and negative feedback loops. *Oncogene*, **24**, 2899–2908.
16. Golden, S.S., Ishiura, M., Johnson, C.H., and Kondo, T. (1997) Cyanobacterial circadian rhythms. *Annu. Rev. Plant Physiol. Plant Mol. Biol.*, **48**, 327–354.
17. Ishiura, M. *et al.* (1998) Expression of a gene cluster *kaiABC* as a circadian feedback process in cyanobacteria. *Science*, **281**, 1519–1523.
18. Kucho, K. *et al.* (2005) Global Analysis of circadian expression in the cyanobacterium *Synechocystis* sp. Strain PCC 6803. *J. Bacteriol.*, **187** (6), 2190–2199.
19. Lemaire, V. *et al.* (2006) Sequential recruitment and combinatorial assembling of multiprotein complexes in transcriptional activation. *Phys. Rev. Lett.*, **96**, 198102.
20. Métivier, R. *et al.* (2003) Estrogen receptor-α directs ordered, cyclical, and combinatorial recruitment of cofactors on a natural target promoter. *Cell*, **115**, 751–763.
21. Blausius, B., Huppert, A., and Stone, L. (1999) Complex dynamics and phase synchronization in spatially extended ecological systems. *Nature*, **399**, 354–359.
22. Hastings, A. and Powell, T. (1991) Chaos in a three-species food chain. *Ecology*, **72** (3), 896–903.
23. Stone, L. and He, D. (2007) Chaotic oscillations and cycles in multi-trophic ecological systems. *J. Theo. Bio.*, **248**, 382–390.
24. Gamarra, J. and Solé, R. (2000) Bifurcations and chaos in ecology: lynx returns revisited. *Ecol. Lett.*, **3**, 114–121.
25. Garcia-Ojalvo, J., Elowitz, M.B., and Strogatz, S.H. (2004) *Proc. Natl. Acad. Sci.*, **101**, 10955–10960.
26. Elowitz, M. and Leibler, S. (2000) A synthetic oscillatory network of transcriptional regulators. *Nature*, **403**, 335–338.
27. Jensen, M.H., Krishna, S., and Pigolotti, S. (2009) The repressor-lattice: feed-back, commensurability, and dynamical frustration. *Phys. Rev. Lett.*, **103**, 118101.

6
Translocation Dynamics and Randomness
Johan Dubbeldam, Vakhtang Rostiashvili, Andrey Milchev, and Thomas Vilgis

6.1
Introduction

Translocation dynamics has attracted a lot of attention recently. In this process, which is ubiquitous in biology, a molecule, initially on one side of the membrane (called the *cis* side) of the cell, moves to the other side (*trans* side), through a nanosized pore (see also Section 6.2). In fact, a biological cell is filled with all kinds of nanopores that control the trafficking of ions and molecules in and out. In this chapter, we review the process of translocation and consider it in the context of complexity and dynamical systems. In translocation, a large number of base pairs interact, finally leading to the emergent behavior in which a chain crosses a membrane. Hence, although the process exhibits a very large number of degrees of freedom, we are actually interested only in one, namely, the fraction of the chain that has passed through the pore at each instant of time. Emergence, being a signature of complexity, is well illustrated by the example of translocation. Other interesting complex behavior in biology is, for example, discussed in Chapter 5. The field of translocation began to develop with experiments that we will describe next.

The pioneering experiments in translocation were carried out by Kasianowicz and coworkers who studied a biological nanopore created by α-hemolysin. This protein creates tiny holes in a membrane 1–2 nm deep. It turned out that DNA is capable of threading such pores, which opened the possibilities of sequencing DNA molecules in an efficient way. The threading was measured by putting a small voltage over the pore, which induces movements of the ions in the solvent, which is then measured as an electric current. As soon as DNA blocks the pore, the current drops, which can then be identified as a translocation event. The duration of the blockage is equal to the time spent by the DNA molecule in the pore and is commonly referred to as the *translocation time* (see Figure 6.1). Only single-stranded DNA chains can find their way through the pore, as the double-stranded DNA is too wide with its 2-nm diameter. Moreover, a single-base resolution, necessary for DNA sequencing, is virtually impossible as the translocation is very fast: a base pair typically spends 1 μs in the pore, which leads to a signal-to-noise ratio too small to ever read off the sequence of

The Complexity of Dynamical Systems. Edited by J. Dubbeldam, K. Green, and D. Lenstra
Copyright © 2011 WILEY-VCH Verlag GmbH & Co. KGaA, Weinheim
ISBN: 978-3-527-40931-0

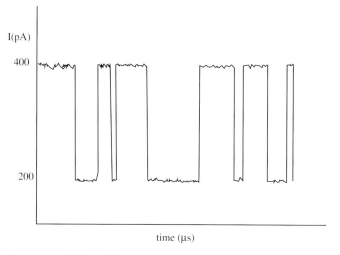

Figure 6.1 Blockage of the pore by a DNA chain causes current drops, allowing measurement of the translocation time distribution.

base pairs, with sufficient degree of statistical significance, from the electric current that is experimentally observed.

Most experiments were carried out for driven DNA chains, as this situation best resembles the situation in real cells *in vivo*, where the *trans* and *cis* sides of the membrane do not generally experience the same electric potential. Moreover, the fact that translocation times for undriven chains greatly exceed those for driven chains leads to further focus on driven chains in experiments. Theoretically, such constraints are not applicable and we therefore start by considering unbiased translocation. Usually, models for unbiased translocation are rather easily generalized to the driven case. Theoretical descriptions of translocation were initiated by Sung and Park [1] and Muthukumar [2]. More recently, a lot of computational and theoretical investigations have been performed [3–6] and characteristic exponents, which determine how the translocation time (τ) scales with the length of the molecule (N), have been calculated. Sung and Park suggested that translocation time should grow with N like N^3. They assumed the diffusion constant D to be inversely proportional to the length of the polymer chain. By arguing that the square of the distance traveled during a time interval τ is of the same order as the polymer length squared, they found that $\tau \simeq N^2/D$, with D inversely proportional to N. This produces a scaling of $\tau \propto N^3$. Muthukumar argued that the translocation time $\tau \simeq N^2$, as the friction is concentrated at the pore, rendering D independent of N.

Computer simulation by Kantor and Kardar [7] in two-dimensional systems resulted in scaling of $\tau \simeq N^\alpha$, with $\alpha = 2.5$; that is, somewhere between the predictions made by Sung and Park and Muthukumar. Numerous models have since then been proposed to explain the value of the exponent α. We will next introduce a few of the models and discuss the remaining controversies and possible resolutions. The aim of this chapter is to illustrate the techniques of Fokker–Planck equations and generalizations thereof in translocation dynamics. Both the statistical physics background and the mathematical solution techniques are discussed. Finally, Monte Carlo (MC) simulations are compared with the analytically obtained results.

6.2
Anomalous Diffusion Model

The model proposed by Sung and Park starts from a description in which the chain is locally in equilibrium with its environment at all times. This implies that one can use a free energy description for the chain, which has one monomer residing in the pore. The typical configuration is schematically depicted in Figure 6.2.

The total free energy (F) consists of two contributions: a contribution of the piece of the chain of length n on the *cis* side of the membrane and the remaining part of length $N-n$ still being on the *trans* side. Hence,

$$F = \frac{1}{2} k_B T \ln[(N-n)n], \tag{6.1}$$

which is maximal at $n = N/2$. The factor $1/2$ in Equation 6.1 stems from the steric hindrance for phantom (not self-avoiding) chains caused by the wall. This implies that a translocation process can be described as Kramer's process in which a free energy barrier has to be overcome, or equivalently one can solve the Fokker–Planck equation associated with this process, which is given by

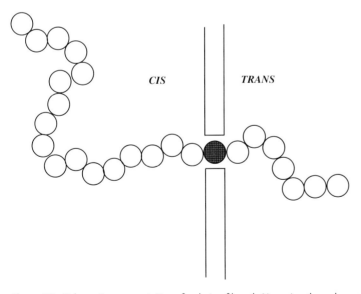

Figure 6.2 Schematic representation of a chain of length N moving through a nanopore. In this case, $n = 7$ beads have moved from the *cis* to the *trans* side.

$$\frac{\partial P(n,t)}{\partial t} = D\frac{\partial^2 P(n,t)}{\partial n^2} + \frac{D}{2}\frac{\partial}{\partial n}\left(\frac{(N-2n)}{(N-n)n}P(n,t)\right), \tag{6.2}$$

which is supplemented by adsorbing boundary conditions in $n = N$ and a reflecting boundary condition at $n = 0$. From Equation 6.2, one can easily find that the translocation time τ is given as $\tau \sim (N^2/D)$. Chuang et al. [8] already noticed that the Fokker–Planck equation (6.2) is, in fact, invariant with respect to the transformation $t \to tN^2$, $n \to nN$, demonstrating that if the translocation process is described by a Fokker–Planck equation with diffusivity D independent of n, the translocation time should scale with N^2. This remains true if instead of phantom chains, self-avoiding chains are considered, as only the factor $1/2$ in Equation 6.2 is changed in that case. However, as noticed by Chuang et al., the Rouse time scales with polymer length as $N^{2\nu+1}$, which in 3d is $N^{2.2}$. This implies that the Rouse time of the chain, which determines the time necessary for the chain to relax to equilibrium, exceeds the translocation time. This conclusion renders the whole analysis in Ref. [2] invalid, as it was based on the presumption of thermodynamic equilibrium.

One should keep in mind that the derivation of the Fokker–Planck equation itself was also reached on the premise that the chain threading through the pore is at all times in thermodynamic equilibrium. From numerical findings in two dimensions using a Monte Carlo algorithm [8], this proved invalid. This was later corroborated by Slater and Gauthier [6]. In the next section, we discuss a possible resolution of the problem in terms of folds. First, we discuss unbiased translocation and then we turn to driven translocation.

6.2.1
Unbiased Translocation

As the assumption of thermodynamic equilibrium of the chain during the threading process is contradicted by numerical simulations, we will instead focus on parts of the chain. Here, we follow Grosberg et al. [9], who put forward the idea not to focus on the entire chain but rather restrict to so-called folds. Their motivation was the analogy with a rope lying on the table and has some folds in it. Just as pulling this rope will cause these folds to straighten up, instead of the entire rope to move, we expect folds in polymers to be able to thread through a pore. This suggests that the folds are the building blocks one should work with. Moreover, as these folds have a much smaller length than the whole chain, these folds can be treated as being in thermal equilibrium. A sketch of a polymer with folds translocating through a nanopore is given in Figure 6.3.

If the *trans* part of a fold has length n, then the corresponding free energy function $F^t(n)/T = -n\ln\kappa - (\gamma_1 - 1)\ln n$, where κ is the connective constant and γ_1 is the surface entropic exponent accounting for the steric hindrance by the wall; $\gamma_1 \approx 0.68$ [10]. For the *cis* part of the fold, one has $F^c(n)/k_B T = -(s-n)\ln\kappa - (\gamma_1 - 1)\ln(s-n)$ so that the total free energy (F) is

$$F(n)/k_B T = -s\ln\kappa - (\gamma_1 - 1)\ln[n(s-n)]. \tag{6.3}$$

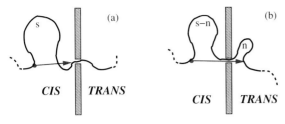

Figure 6.3 Schematic representation of a chain of length N showing a fold of length s. While translocating, the fold decomposes into two pieces. Initially, the fold is on the *cis* side of the membrane (a). Threading causes fold fragmentation (b).

One can, therefore, ascribe to the fold *cis–trans* transition a rather broad barrier given by $F(n)$. The corresponding activation energy of the fold can be calculated as

$$\Delta E(s) = F(s/2) - F(1) = (1-\gamma_1) k_B T \ln s. \tag{6.4}$$

In order to find a characteristic time for the translocation process, we consider it as being an activated process of the polymer over an energy barrier that in our case is given by Equation 6.4 (see Figure 6.4). The time span for a certain part of the fold (s) to travel across the pore is then given by

$$t(s) = t_R(s) \exp(\Delta E(s)) \sim s^{2\nu + 2 - \gamma_1}, \tag{6.5}$$

where $t_R(s)$ denotes the Rouse time of the polymer, that is, the time it would take the polymer to move from *cis* to the *trans* side without the presence of a membrane, which

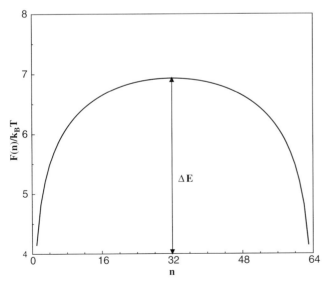

Figure 6.4 The free energy barrier that has to be overcome by the polymer chain while crossing the membrane from *cis* to *trans* side.

is known to obey $t_R(s) = s^{2\nu+1}$, with ν the Flory exponent, accounting for excluded volume effects. We can invert this relation that yields

$$\langle s^2 \rangle \propto t^{2/(2\nu+2-\gamma_1)}. \tag{6.6}$$

Equation 6.6 implies that the second moment of s does not scale linearly with t, as in the case of ordinary diffusion, but rather grows with time with an anomalous exponent, signaling anomalous diffusion. By putting $t(s)$ equal to the translocation time τ and $s = N$, we find that $\tau \sim N^{2/\alpha}$ with $\alpha = 2/(2\nu+2-\gamma_1)$. Putting in values for the constants, we find that for $d = 3$, $\nu = 0.588$, $\gamma_1 = 0.68$, leading to $\alpha = 0.801$. Surprisingly for $d = 2$, we find $\alpha = 0.783$, although $\nu = 0.75$ and $\gamma_1 = 0.945$ in two dimensions. The small differences between the 2d [8] and 3d [5] values for α may explain why the values of the exponents found in numerical experiments in 2d and 3d are so close.

6.2.1.1 Continuous Time Random Walks and Fractional Derivatives

A possible explanation of the anomalous diffusion behavior that arose in Equation 6.6 is that the translocation process in which a monomer moves from one side of the membrane to the other side during a certain amount of time cannot be described by a simple random walk (SRW) process but instead executes the so-called continuous time random walk (CTRW). In a CTRW, a walker on a one-dimensional lattice steps to the left or right, according to a certain waiting time distribution. This is in contrast to the SRW, which does not pertain to any nontrivial waiting time distribution, rather each time, a step is made either to the left or to the right. A CTRW is, therefore, much more general and depending on its waiting time distribution it can describe processes varying from an SRW to the so-called Lévy flights. One can show that a Poissonian waiting time distribution leads to an SRW [11]. A long tail in the waiting time distribution, which decays with time as $t^{-\beta}$, with $0 < \beta < 1$, leads to a stochastic process that is qualitatively different from an SRW. In particular, it leads to a probability distribution function for the first passage time that has a diverging first moment. We will not go into further detail, but direct the reader to the review paper by Metzler and Klafter [11]. In this paper, the authors demonstrate how a CTRW leads to modifications in the Fokker–Planck equation, when the CTRW process is described in continuous spatial and time coordinates. It turns out that the waiting time distribution leads to fractional derivatives entering the Fokker–Planck equation.

Before embarking on the fractional Fokker–Planck equation (FFPE), we stress that a CTRW and the corresponding FFPE *do not necessarily follow* from the observation that the diffusion is anomalous. Other processes could also account for such subdiffusive behavior. Our choice for using a FFPE is mainly motivated by the fact that its results are in good correspondence with numerical experiments, and it allows generalization of the translocation process to folds.

Another reason to use a CTRW description is that a threading chain is expected to feel its environment, which provides it with a certain memory. Although such effects are not incorporated in our model, in realistic circumstances a biological pore such as the one induced by α-hemolysin is not perfectly smooth, but very narrow and irregular. Consequently, a monomer inside the pore feels the environment through interactions with irregularities of the pore, which may lead to a broad distribution of waiting times. If this distribution is such that it depends nontrivially on time, which is the case for translocating a polymer chain that is not in thermal equilibrium and therefore does not satisfy detailed balance [11, 12], the threading process is expected to be well described by a CTRW. In fact, it has been observed in recent experiments [13] that the time distribution of translocating single-stranded DNA (ssDNA) is very wide with a fat tail for long threading times, which suggests that the CTRW formalism indeed provides the right framework to properly treat translocation through real pores. Although such effects are precluded from our model that assumes a perfectly smooth pore, the description by the FFPE is capable of including such a generalization.

In the framework of anomalous diffusion, the translocation process is governed by the equivalent of the diffusion equation for SRW: the fractional diffusion equation (FDE), which reads

$$\frac{\partial}{\partial t} P(s,t) = {}_0 D_t^{1-\alpha} K_\alpha \frac{\partial^2}{\partial s^2} P(s,t), \tag{6.7}$$

where K_α is the fractional diffusion coefficient and $P(s,t)$ denotes the probability of having the segment s at time t in the pore. The operator ${}_0 D_t^{1-\alpha}$, commonly referred to as the Riemann–Liouville fractional derivative, is defined as

$$ {}_0 D_t^{1-\alpha} P(s,t) = (1/\Gamma(\alpha))(\partial/\partial t) \int_0^t dt' P(s,t')/(t-t')^{1-\alpha}, \tag{6.8}$$

where $\Gamma(\alpha)$ is the gamma function. The memory effects enter Equation 6.7 via the Riemann–Liouville derivative (see Ref. [14] for a detailed account of this operator).

Equation 6.7 is supplemented by a reflecting boundary condition in $s = 0$: $(\partial/\partial s) P(0,t) = 0$, which prevents the chain from leaving the pore on the *cis* side and an absorbing one in $s = N$: $P(s = N, t) = 0$. The initial condition is given by the usual equation:

$$P(s, t = 0) = \delta(s - s_0). \tag{6.9}$$

We can next solve Equation 6.7 by using separation in s and t, that is, we expand the solution as

$$P(s,t) = \sum_{n=0}^{\infty} \phi_n(s) T_n(t), \tag{6.10}$$

where $\phi_n(s)$ satisfies

$$K_\alpha (d^2/ds^2) \phi_n(s) + \lambda_{n,\alpha} \phi_n(s) = 0 \tag{6.11}$$

and the temporal part $T_n(t)$ obeys

$$(d/dt) T_n(t) = -\lambda_{n,\alpha} \, {}_0 D_t^{1-\alpha} T_n(t). \tag{6.12}$$

The solution of Equation 6.12 is given in terms of Mittag–Leffler functions $E_\alpha(x)$, which are the generalization of the exponential function $\exp(x)$, and defined as a power series by

$$E_\alpha(x) = \sum_{n=0}^{\infty} x^k / \Gamma(1 + \alpha k) \tag{6.13}$$

For $\alpha = 1$, this function reduces to the ordinary exponential function. The complete solution of Equation 6.7 is therefore given by

$$P(s,t) = \frac{2}{N} \sum_{n=0}^{\infty} \cos\left[\frac{(2n+1)\pi s_0}{2N}\right] \cos\left[\frac{(2n+1)\pi s}{2N}\right] \\ \times E_\alpha\left[-\frac{(2n+1)^2 \pi^2}{4N^2} K_\alpha t^\alpha\right]. \tag{6.14}$$

Mean First Passage Time In experiments, the probability distribution (6.14) is not accessible, but rather the translocation time is, as we explained in Section 6.1. This time can be identified as the first passage time associated with (6.14). The first passage times are distributed according to $Q_{s_0}(t) = -(d/dt) \int_0^N P(s,t) ds$ [15], where the subscript s_0 indicates that the mean passage distribution depends, of course, on the starting position s_0. The expression for $Q_{s_0}(t)$ reads

$$Q_{s_0}(t) = \frac{\pi K_\alpha t^{\alpha-1}}{N^2} \sum_{n=0}^{\infty} (-1)^n (2n+1) \cos\left[\frac{(2n+1)\pi s_0}{2N}\right] \\ \times E_{\alpha,\alpha}\left[-\frac{(2n+1)^2 \pi^2}{4N^2} K_\alpha t^\alpha\right], \tag{6.15}$$

where we introduced the generalized Mittag–Leffler functions $E_{\alpha,\alpha}(x)$, which are defined in terms of the power series

$$E_{\alpha,\alpha}(x) = \sum_{k=0}^{\infty} x^k / \Gamma(\alpha + k\alpha). \tag{6.16}$$

For short times, the $Q_{s_0}(t)$ depends on the initial condition s_0 and for long times this dependence vanishes. As we are particularly interested in the long time tails, we approximate the Mittag–Leffler functions by their asymptotic expansions

$$E_\alpha(-\lambda_{n,\alpha} t^\alpha) \propto 1/\Gamma(1-\alpha) \lambda_{n,\alpha} t^\alpha \tag{6.17}$$

and

$$E_{\alpha,\alpha}(-\lambda_{n,\alpha} t^\alpha) \propto \alpha/\Gamma(1-\alpha) \lambda_{n,\alpha}^2 t^{2\alpha}, \tag{6.18}$$

from which we find that the long time tail of $Q_{s_0}(t)$, denoted by $Q(t)$, is given by

$$Q(t) \propto \alpha N^2 / 2\Gamma(1-\alpha) K_\alpha t^{1+\alpha}. \tag{6.19}$$

We note that the first moment of $Q(t)$ corresponding to the average translocation time $\tau = \langle t \rangle = \int_0^\infty t Q(t) dt$ diverges. However, in a laboratory experiment there always exists some upper time limit t^*. Taking this into account, one can show that an "experimental" first passage time scales as $\tau \sim N^{2/\alpha}$ [9], which we observe in our MC simulation.

Moments $\langle s \rangle$ and $\langle s^2 \rangle$ versus Time Besides the mean first passage time distribution (6.15), the moments of the average value of the translocation coordinate $s(t)$ and its variation $\langle s^2(t) \rangle$ can be used to characterize the threading process. From the probability distribution, one easily finds that $\langle s(t) \rangle$ can be expressed in terms of Mittag–Leffler functions as

$$\frac{\langle s \rangle(t)}{N} = 1 - \frac{2 \sum_{n=0}^{\infty} \frac{1}{(2n+1)^2} E_\alpha\left[-\frac{(2n+1)^2 \pi^2}{4N^2} K_\alpha t^\alpha\right]}{\pi \sum_{n=0}^{\infty} \frac{(-1)^n}{(2n+1)} E_\alpha\left[-\frac{(2n+1)^2 \pi^2}{4N^2} K_\alpha t^\alpha\right]} \tag{6.20}$$

Since $E_\alpha(t=0) = 1$, the initial value $\langle s \rangle(t=0) = 0$ (we put $s_0 = 0$) as it should be. In the opposite limit, $\langle s \rangle(t \to \infty) = N/3$, that is, the function goes to a *plateau*. The result for the second moment, $\langle s^2 \rangle = \int_0^N s^2 W(s,t) ds / \int_0^N P(s,t) ds$, can be cast in the following form:

$$\frac{\langle s^2 \rangle(t)}{N^2} = 1 - \frac{8 \sum_{n=0}^{\infty} \frac{(-1)^n}{(2n+1)^3} E_\alpha\left[-\frac{(2n+1)^2 \pi^2}{4N^2} K_\alpha t^\alpha\right]}{\pi^2 \sum_{n=0}^{\infty} \frac{(-1)^n}{(2n+1)} E_\alpha\left[-\frac{(2n+1)^2 \pi^2}{4N^2} K_\alpha t^\alpha\right]}. \tag{6.21}$$

Again, it can be readily shown that $\langle s^2 \rangle(0) - \langle s \rangle^2(0) = 0$, whereas in the long time limit $\langle s^2 \rangle(t \to \infty) - \langle s \rangle^2(t \to \infty) = N^2/9$. A calculation for the ordinary Fokker–Planck equation shows that these plateaus also occur in the ordinary Fokker–Planck equation subject to the same boundary conditions. Therefore, they do not dependent on memory effects, but rather on intrinsic properties of the Fokker–Planck equation with the boundary conditions imposed in general.

6.2.2
Monte Carlo Simulations

In order to check our analytical calculations, we performed dynamic MC simulations using the algorithm used by Milchev et al. [16]. The MC computations describe N beads, connected by finitely extensible nonlinear elastic (FENE) springs, whose potential is given by

$$U_{\text{FENE}} = -\frac{KR^2}{2} \ln\left[1 - \frac{(l-l_0)^2}{R^2}\right]. \tag{6.22}$$

Here, l is the length of an effective bond, which can vary between l_{min} and l_{max}, with $R = l_{max} - l_0$, and l_0 the equilibrium value for which the potential is minimal. We chose $l_{max} = 1, l_{min} = 0.4, l_0 = 0.7$. The spring constant K was taken equal to $40 k_B T$. The nonbonded beads interact by a Morse potential U_M given as

$$U_M = \varepsilon_M [\exp(-2\alpha(r-r_{min})) - 2\exp(-\alpha(r-r_{min}))], \quad (6.23)$$

where r is the distance between the two beads, and we have chosen $r_{min} = 0.8$ and $\varepsilon_M = 1$. The value of $\alpha = 24$, which allows to use an efficient link cell algorithm [16].

An elementary Monte Carlo move consists of picking a bead at random and trying to displace it to a new position at random. Such a move is accepted with a probability given by the Metropolis algorithm. It is well known that such an algorithm, based on local moves, faithfully represents Rouse dynamics for the polymer chain.

We have carried out simulations for different chain lengths varying from $N = 16$ to $N = 256$. Each runs starts with one monomer on the *trans* side and complete retraction of the chain to the *cis* side is prohibited. During the runs we record the translocation coordinate $s(t)$ and the translocation time τ. For each chain length, we average over 10^4 runs, to obtain statistically good averages.

Comparison Theory and MC Simulations In Figure 6.5a, we show a master plot of the translocation time distribution $Q(t)$ derived from Equation 6.15 for different chain lengths $N = 16, 32, 64, 128, 256$. For the calculation of data, we have used Mathematica with a special package for computation of Mittag–Leffler functions [17]. Evidently, all curves collapse on a single one when time is scaled to $t \propto N^{2/\alpha}$ with the predicted $\alpha = 0.8$.

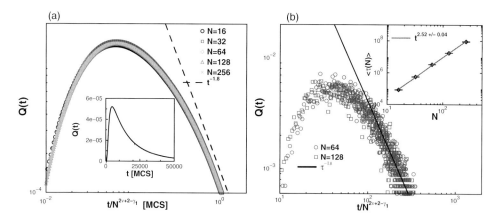

Figure 6.5 The translocation time distribution function $Q(t)$. (a) Scaling plot of the theoretical predictions calculated from Equation 6.15 for different chain lengths N. Dashed line denotes the long time asymptotic tail with slope -1.8. The inset shows $Q(t)$ for $N = 256$ in normal coordinates. (b) The FPDT $Q(t)$ from the MC simulation for $N = 64, 128$. The inset shows the expected $\langle \tau \rangle$ versus chain length N dependence and the straight line is a best fit with slope $\approx 2.52 \pm 0.04$.

The long time tail for this value of α should exhibit a slope of −1.8. The inset in Figure 6.5a reveals the long tail of $Q(t)$ for large times. A comparison with Figure 6.5b demonstrates good agreement with the simulation data despite some scatter in the FPDF even after averaging over 10 000 runs. Distinguishing an exponential decaying probability distribution from a power law distribution is notoriously hard. Even varying the bin size with translocation time could not resolve the exact shape of the tail of the distribution. As shown in the inset, the mean translocation time scales as $\langle \tau \rangle \propto N^{2.5}$ in good agreement with the predicted α = 0.8. These MC results therefore at least do not contradict the FFPE description that we presumed initially.

Besides the mean first passage time distribution, the moments contain a lot of information of the translocation process. We therefore recorded the first- and second-order moments of $\langle s(t) \rangle$ during the MC simulations. The results are shown in Figure 6.6. One clearly sees the appearance of a plateau for large t.

An inspection of Figure 6.6, where the time variations of the PDF $P(s, t)$ moments are compared, demonstrates again that data from the numeric experiment and the analytic theory agree well within the limits of statistical accuracy (which is worse for $N = 256$). Not surprisingly, the timescale of the MC results does not coincide with that of Equations 6.20 and 6.21 since in the latter we have set K_α, which fixes the timescale, equal to unity.

In summary we have shown [18] that the translocation dynamics of a polymer chain threading through a nanopore might be anomalous in its nature. We have succeeded to calculate the anomalous exponent $\alpha = 2/(2\nu + 2 - \gamma_1)$ from simple scaling arguments and embedded it in the fractional diffusion formalism. We derived exact analytic expressions both for the translocation time probability distribution and for the moments of the translocation coordinates that are shown to agree well with

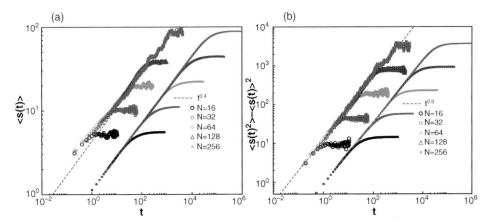

Figure 6.6 Variation of the first and second moments of the PDF $P(s, t)$ with time for chain lengths $N = 16, 32, 64, 128, 256$. (a) Log–log plot of the first moment $\langle s \rangle$ versus time t from a MC simulation (big symbols) and from Equation 6.20 (small symbols). The dashed line denotes $t^{\alpha/2}$ with a slope of 0.4. (b) The same as in (a) but for $\langle s^2 \rangle - \langle s \rangle^2$. Analytical data is obtained from Equation 6.21. The dashed line has a slope of 0.8.

our MC simulation data. The present treatment can be readily generalized to account for a driving force on the chain and results for this case will be reported in the next section.

6.2.3
Driven Translocation

As mentioned in Section 6.1, most translocation experiments, as well as processes, in living cells are performed under circumstances in which there is a potential difference between the inside and the outside environment of the cell. Therefore, we consider here the problem of translocation in the presence of external driving force that in most cases consists of an electrical field. The presence of the external force imposed on the translocating chain leads to a nonisotropic *cis–trans* transition of the folds [19]. It can be quantified within the FFPE formalism [11]. In our case, FFPE with a driving potential $U(s)$ has the form

$$\frac{\partial}{\partial t} P(s,t) = {}_0 D_t^{1-\alpha} \left[\frac{\partial}{\partial s} \frac{U'(s)}{\xi_\alpha} + K_\alpha \frac{\partial^2}{\partial s^2} \right] P(s,t). \tag{6.24}$$

The meaning of the symbols is as explained in Equation 6.7. The external field $U(s)$ in Equation 6.24 is a simple linear function of the translocation s coordinate, namely, $U(s) = -\Delta\mu\, s$, where $\Delta\mu = \mu_1 - \mu_2$. μ_1 and μ_2 can be thought of as chemical potentials on the *cis* and *trans* sides of the membrane or $\Delta\mu$ can be considered proportional to the applied voltage difference over the membrane. As before, we supply boundary conditions that are reflecting in $s = 0$ and absorbing in $s = N$. The initial distribution is concentrated in s_0, that is, $P(s, t = 0) = \delta(s-s_0)$, as before.

Following the same strategy as for the unbiased case, we again try to expand the solution of Equation 6.24 in terms of $T_n(t)$ and $\psi_n(s)$ (see also Risken [15])

$$P(s,t) = \exp(\Phi(s_0) - \Phi(s)) \sum_{n=0}^{\infty} T_n(t) \psi_n(s) \psi_n(s_0), \tag{6.25}$$

where we introduced $\Phi(s) = U(s)/2k_B T$. The functions $\psi_n(s) = \exp(\Phi(s))\varphi_n(s)$, where the $\varphi_n(s)$ obey the equations

$$[(d^2/ds^2) - f(d/ds) + \lambda_{n,\alpha}/K_\alpha] \varphi_n(s) = 0. \tag{6.26}$$

Here, the dimensionless force f is defined by $f \equiv \Delta\mu/k_B T$. The eigenvalues $\lambda_{n,\alpha}$ can be readily found by applying the boundary conditions to eigenfunctions φ_n. For the temporal part, we again find

$$(d/dt) T_n(t) = -\lambda_{n,\alpha}\, {}_0 D_t^{1-\alpha} T_n(t), \tag{6.27}$$

which can be solved in terms of the Mittag–Leffler functions.

The important change with respect to the unbiased case is in the equation for eigenvalues, which is transcendental and given by

$$-2\sqrt{\kappa_n/f} = \tan(\sqrt{\kappa_n} N), \tag{6.28}$$

where $\kappa_n = \lambda_{n,\alpha}/K_\alpha - f^2/4$. This equation can in general be solved graphically or numerically. In the limiting case that $fN \ll 1$, the κ_n can be calculated exactly as $\kappa_n = \lambda_{n,\alpha}/K_\alpha = (2n+1)^2\pi^2/4N^2$ and the eigenfunctions take on the form $\varphi_n(s) = \sqrt{2/N}\cos[(2n+1)\pi s/2N]$. The resulting solution for $P(s,t)$ at $f=0$ reduces to that of the force-free case as was considered in the previous section.

In the opposite limit of a large driving force, $fN \gg 1$, the eigenvalue spectrum reads

$$\lambda_{n,\alpha} = (f^2/4 + n^2\pi^2/N^2)K_\alpha \tag{6.29}$$

and the eigenfunctions $\psi_n(s) = \sqrt{2/N}\sin(n\pi s/N)$, so that the resulting solution becomes

$$P(s,t) = \frac{2}{N}e^{f(s-s_0)/2}\sum_{n=0}^{\infty}\sin\left[\frac{n\pi s_0}{N}\right]\sin\left[\frac{n\pi s}{N}\right]$$

$$= \times E_\alpha\left[-\left(\frac{f^2}{4}+\frac{n^2\pi^2}{N^2}\right)K_\alpha t^\alpha\right]. \tag{6.30}$$

In the limit of strong driving force, the translocation times are relatively (compared to the force-free case) short and we could use the small argument approximation for the Mittag–Leffler function $E_\alpha(-x)$, that is, $E_\alpha(-x) \simeq \exp[-x/\Gamma(1+\alpha)]$ at $x \ll 1$. This makes it possible to obtain an explicit analytical expression for $P(s,t)$ that can be derived by replacing the summation with an integral in Equation 6.30. In doing so one should use the relation $2\sin(n\pi s_0/N)\sin(n\pi s/N) = \cos[n\pi(s-s_0)/N] - \cos[n\pi(s+s_0)/N]$. Then, one can integrate over n explicitly, taking the limit $s_0 \to 0$, and finally normalize the FPTD: $p(s,t) \equiv \lim_{s_0 \to 0} P(s,t)/\int_0^N P(s,t)ds$. This eventually yields

$$p(s,t) = \frac{\exp\{-(s-f\tilde{t})^2/4\tilde{t}\}}{\sqrt{\pi\tilde{t}}\{\mathrm{erf}[f\sqrt{\tilde{t}}/2] - \mathrm{erf}[(f\tilde{t}-N)/2\sqrt{\tilde{t}}]\}}, \tag{6.31}$$

where the dimensionless force $f = \Delta\mu/T$, $\tilde{t} = K_\alpha t^\alpha/\Gamma(1+\alpha)$, and $\mathrm{erf}(x)$ is the error function. Our further theoretical findings are mainly based on Equations 6.30 and 6.31 for PDF.

In the chain translocation experiment, the initial position s_0 can be fixed and the distribution of the translocation times is actually equivalent to the *first passage time distribution* (FPTD) $Q(s_0,t)$ [15]. The relation $Q(s_0,t) = -(d/dt)\int_0^N P(s,t)ds$ [15] enables to calculate FPTD explicitly. Starting from Equation 6.30, we arrive at the expression

$$Q(s_0,t) = \frac{\pi K_\alpha e^{f(N-s_0)/2}}{N^2 t^{1-\alpha}}\sum_{n=0}^{\infty}(-1)^{(n-1)}\sin\left(\frac{n\pi s_0}{N}\right)$$

$$\times E_{\alpha,\alpha}\left[-\left(\frac{f^2}{4}+\frac{n^2\pi^2}{N^2}\right)K_\alpha t^\alpha\right], \tag{6.32}$$

where the generalized Mittag–Leffler function $E_{\alpha,\alpha}(x) = \sum_{k=0}^{\infty} x^k/\Gamma(\alpha+k\alpha)$.

In the same manner as above, we could use the small argument approximation for the generalized Mittag–Leffler function $E_{\alpha,\alpha}(-x)$, that is, $E_{\alpha,\alpha}(-x) \simeq (\alpha/\Gamma(1+\alpha)) \exp[-x/\Gamma(1+\alpha)]$ at $x \ll 1$, to obtain the explicit analytical expression for $Q(s_0, t)$. The substitution of the summation by integration in Equation 6.32 and the use of the relation $(-1)^{n-1} \sin(n\pi s_0/N) = -\cos(n\pi)\sin(n\pi s_0/N) = \sin[n\pi(1-s_0/N)] - \sin[n\pi(1+s_0/N)]$ enable us to finally obtain the normalized FPTD, $\lim_{s_0 \to 0} Q(s_0, t)/\int Q(s_0, t)dt \to Q(t)$. It is given by the following expression:

$$Q(t) = \frac{\alpha}{4\pi^{1/2} f\tilde{t}} \left[\frac{\Gamma(1+\alpha)}{K_\alpha t^\alpha}\right]^{1/2} \left[\frac{N^2 \Gamma(1+\alpha)}{K_\alpha t^\alpha} - 2\right]$$

$$\times \exp\left\{-\frac{[N-f(K_\alpha t^\alpha)/\Gamma(1+\alpha)]^2}{4(K_\alpha t^\alpha)/\Gamma(1+\alpha)}\right\}. \quad (6.33)$$

As one can see, after normalization the dependence on the initial value $s_0 \to 0$ drops. It is of interest that FPDT given by Equation 6.33 exactly coincides (at $\alpha = 1$, that is, in the Brownian dynamics limit) with the corresponding expression in the paper by Lubensky and Nelson [20]. It is also evident from Equation 6.33 that the maximum position scales as $t_{\max} \propto (N/f)^{1/\alpha} = (N/f)^{1.25}$. Nevertheless, the function $Q(t)$ is quite skewed and we will see below that the average translocation time $\tau = \int tQ(t)dt$ (which is presumably measured in an experiment) scales differently.

6.3
Statistical Moments $\langle s \rangle$ and $\langle s^2 \rangle$ versus Time

We now turn to the calculation of the average translocated length $\langle s \rangle$ and its variation given by $\langle s^2 \rangle$. The recording of statistical moments' time dependence, $\langle s(t) \rangle = \int_0^N sw(s,t)ds$ and $\langle s(t)^2 \rangle = \int_0^N s^2 w(s,t)ds$, is very instructive (as in the force-free case [20]) for the consistency check. Starting from Equation 6.31, the calculation of the first moment yields

$$\langle s(t) \rangle = f\tilde{t} + 2\sqrt{\frac{\tilde{t}}{\pi}} \frac{\exp[-f^2\tilde{t}/4] - \exp[-(f\tilde{t}-N)^2/4\tilde{t}]}{\mathrm{erf}[f\sqrt{\tilde{t}}] - \mathrm{erf}[(f\tilde{t}-N)^2/2\sqrt{\tilde{t}}]}. \quad (6.34)$$

It can easily be shown that in the large time limit $\langle s \rangle \to N$. In the same manner, the second moment reads

$$\langle s^2(t) \rangle = f^2\tilde{t}^2 + 2\tilde{t} + 2\sqrt{\frac{\tilde{t}}{\pi}} \frac{f\tilde{t}\exp[-f^2\tilde{t}/4] - (f\tilde{t}+N)\exp[-(f\tilde{t}-N)^2/4\tilde{t}]}{\mathrm{erf}[f\sqrt{\tilde{t}}] - \mathrm{erf}[(f\tilde{t}-N)^2/2\sqrt{\tilde{t}}]}. \quad (6.35)$$

In Equations 6.34 and 6.35, the notations are the same as in Equation 6.31. The detailed check of these relations will be given below. Here, we only note that for large times $1 < \tilde{t} < N/f$, the exponential terms in Equations 6.34 and 6.35 vanish so that to leading order the moments vary as $\langle s(t) \rangle \propto t^{\alpha}$ and $\langle s(t)^2 \rangle \propto t^{2\alpha}$. Again, it can be shown that at $t \to \infty$ the moments $\langle s(t) \rangle$ and $\langle s^2(t) \rangle$ saturate to plateaus that scale like N and N^2, respectively, as they should.

Now that we have derived the first and second moments as well as the mean first passage time distribution, we try to find scaling arguments to corroborate our findings.

Scaling Arguments As a natural scaling variable, we take $|\langle U(s) \rangle| = fN$, on which all physical relevant quantities will depend. Next, we consider the translocation time τ. We presume that the driven translocation rate scales as

$$\tau^{-1} = \tau_0^{-1} \phi(fN), \qquad (6.36)$$

where $\tau_0 \propto N^{2\nu + 2 - \gamma_1}$ denotes the translocation time in the force-free case.

The scaling function $\phi(x)$ behaves in the following way: $\phi(x \ll 1) \simeq 1$ and $\phi(x \gg 1) \simeq x$ because at $fN \gg 1$ we would expect that the translocation rate is proportional to the force f. As a result, we come to the conclusion that at $fN \gg 1$ the translocation time scales as

$$\tau \propto \frac{1}{f} N^{2\nu + 1 - \gamma_1} \qquad (6.37)$$

Taking into account the values for ν and γ_1 given above, we arrive at the following estimations: at $d = 3$ the translocation exponent $\theta = 2\nu + 1 - \gamma_1 = 1.496$ and at $d = 2$ the exponent $\theta = 2\nu + 1 - \gamma_1 = 1.56$. This is quite close to the estimation given by Kantor and Kardar [7], $\theta = 1.53$.

Comparison of Monte Carlo Simulation versus Theory (Driven Case) Again, we perform Monte Carlo simulations; this time the bead in the pore gets assigned an additional energy, corresponding to the force times the displacement. All data were averaged over 10^4 runs in order to minimize the scatter. In Figure 6.7, we show the PDF $Q(\tau)$ of a polymer chain with $N = 128$ for three different values of the driving force, $f = 0.5, 0.8$, and 1.0. Although the MC data are somewhat scattered, especially for $f = 0.5$, the agreement with the analytic expression, Equation 6.33, is very good. Since we set the generalized diffusion coefficient $K_\alpha \equiv 1$ and $\Gamma(1 + \alpha) \approx 0.931$ for $\alpha = 0.8$, the comparison with MC results suggests that a time unit in the FFPE corresponds roughly to 500 MCS.

Using the PDF $Q(\tau)$, one may determine the MFPT (or, translocation times) τ that are compared in Figure 6.8 for $16 < N < 512$ and six values of the driving force f. Evidently, for both theory and simulation the data collapse on the master curves $f\tau \propto N^{1.5}$ if one scales τ with the respective force (cf. Equation 6.37). It is seen that the simulation data are shifted up by a factor of approximately 500, which translates the MC time into conventional time units. The variation of the moments $\langle s \rangle$ and $\langle s^2 \rangle$ is displayed in Figure 6.9. Again, a perfect collapse of the transients is achieved by scaling the time with the applied force $t \to ft$. As mentioned above, this course is very well accounted for by Equations 6.34 and 6.35. Thus, for $ft \ll 1$ one can readily obtain

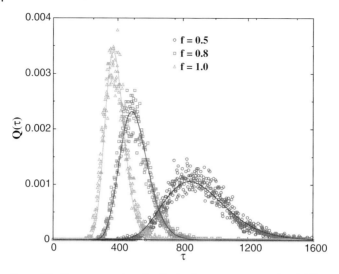

Figure 6.7 First passage time distribution functions at $N = 128$ and different forces as calculated from MC data (symbols) and the theoretical prediction (Equation 6.33; solid lines).

from Equation 6.34 as a leading term $\langle s \rangle \propto t^{\alpha/2}$ while for $1 < ft < N$ one has $\langle s \rangle \propto t^{\alpha}$. As indicated in Figure 6.9, the observed agreement between theory and computer experiment is remarkable indeed. Notably this finding suggests that even the presence of a driving force *does not* eliminate the anomalous character of the

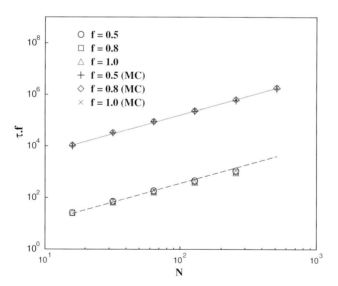

Figure 6.8 The average translocation time versus chain lengths. The upper line represents the results of MC simulation, the lower refers to the theoretical prediction obtained by the proper numerical integration of FPTD (Equation 6.33). Both dashed lines correspond to a power-law dependence with exponent 1.5.

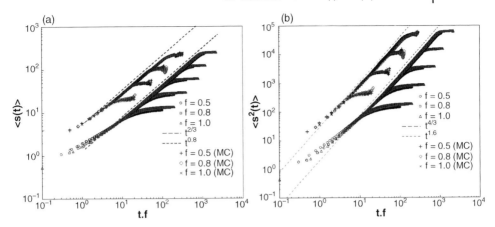

Figure 6.9 Statistical moments versus reduced time $f \cdot t$ from MC data and from the analytical results, Equations 6.34 and 6.35 for chain length $16 \leq N \leq 256$. (a) The first moment $\langle s(t) \rangle$: a short dashed line denotes $\langle s \rangle \propto t^{\alpha}$. (b) The second moment $\langle s^2(t) \rangle$: the short dashed line denotes $\langle s^2 \rangle \propto t^{2\alpha}$.

translocation process contrary to what one might intuitively expect. This result thus resolves a problem, raised initially by Metzler and Klafter [12]. The universal exponent $\alpha = 2/(2\nu + 2 - \gamma_1)$ for unbiased threading through a pore is not suppressed by the driving force. One may thus conclude that the measurement of the number of translocated segments with time could provide a means for direct observation of anomalous diffusion.

6.3.1
Translocation: A Quasi-Equilibrium Process?

We have so far seen that translocating a polymer chain from the *cis* to the *trans* side of a membrane can be described by a fractional Fokker–Planck equation. Moreover, by using the fold picture, we were able to relax the conditions on the chain for the thermodynamic equilibrium assumption to hold. Of course, introducing a sufficiently strong driving force certainly drives a chain through a pore much faster than diffusion. This will eventually lead to a translocation time that is shorter than the equilibration time of the folds. This means that for large enough driving forces even the folds will no longer be in quasi-thermal equilibrium. Nevertheless, the description in terms of a FFPE gives surprisingly accurate results that compare extremely well with MC simulations. It seems that there is still an open problem here and we need to scrutinize the thermodynamic equilibrium assumption in more detail to explain why the FFPE works so surprisingly well. A recent work by Gauthier and Slater [6] using MD simulation for very short chains ($N \approx 30$) has shown that when translocating a polymer chain without a driving force and starting with the middle of the chain in the pore eventually leads to accelerated motion of the chain. This acceleration, which is entropically driven, leads to a radius of gyration that increases

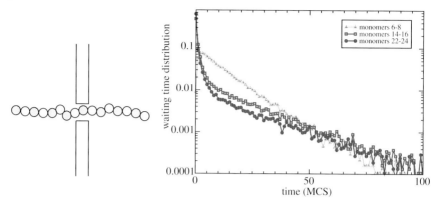

Figure 6.10 When a hashed bead passes through the pore, the time till the next hashed bead crosses the pore for the first time is recorded. This is done for beads 6–8 (triangles upper curve), beads 16–18 (squares middle curve), and beads 22–24 (circles lowest curve).

drastically during the last stage of the translocation process. This further indicates that the process is not in quasi-equilibrium. Moreover, they found that equilibrium translocation, which can be established by artificially equilibrating the chain, speeds up the translocation process [6]. To gain further insight into these issues, we have performed numerical simulations in which we let a chain of length $N = 64$ diffuse through a pore. The chain is divided into 32 pieces of 2 monomers each, and we record for each piece when it has shifted 2 units to the *trans* side for the first time. From this data, we can extract the velocity of the chain depending on its position s, or equivalently, the time needed for two monomers to travel from the *cis* to the *trans* side.

In Figure 6.10, we plotted for three pairs of monomers the probability distribution of the time it takes for two monomers to have moved two units in the *trans* direction for the first time, which we call the waiting time. We find that this time depends on the position of the monomers (s) in the chain. From Figure 6.10, one immediately observes that the average time interval between two first passage events decreases with s, except for very long times, implying an increasing speed during the translocation process. This is in agreement with the results in Ref. [6]. We like to remark that for the first few monomers, we have, of course, in addition to the entropic effect, the influence of the reflecting boundary condition, which, however, rapidly decays and will no longer be felt by monomers 12 and higher. We can therefore attribute the acceleration witnessed in Figure 6.10 for the curves of momoner pairs 14–16 and 22–24 to the entropic driving. In order to have a consistent picture that takes this nonequilibrium behavior into account, we propose to perform MD simulations, whose results can then be compared with MC simulations. This should help clarify the question whether the assumption of (fold) equilibrium is justified and how a possible nonequilibrium situation can still lead to scaling exponents that are so well described by the FFPE.

6.4
Discussion

We have considered a theoretical model of polymer translocation as an example of a biological system in which macroscopic stochastic behavior emerges from a system (polymer chain) interacting with a lot of components. We have discussed issues that are important for complex systems in general. In particular, the role and validity of quasi-thermodynamic equilibrium (see also Chapter 7) for translocation are questioned and examined. Understanding the dynamics of the translocation process that drives long molecules, such as DNA, through small pores is of great practical importance. It could lead to an enormous reduction in sequencing time and cost reduction of sequencing [21]. In order for this ultimate goal to be reached, a number of technical and theoretical problems have to be resolved. The influence of roughness of a biological pore on the translocation time is still largely unknown, whereas this is essential in order to make quantitative predictions for polymer translocation. The effect of the cell environment is only just beginning to be explored. The issue of how to increase the resolution of the experiments is another open problem; the same is true for the theoretical question if it would in principle be possible, using longer pores or applying varying electric fields, to construct a system that would be able to distinguish between the different sequences of C,G,T,A nucleotides in DNA.

In conclusion, we may say that translocations constitute an interesting example of a dynamical system, showing a number of features inherent to complex systems. As with other complex systems, a lot of challenges lie ahead, whose resolution could bring fast and cheap sequencing closer to us in the future.

References

1. Sung, W. and Park, P.J. (1996) *Phys. Rev. Lett.*, **77**, 783.
2. Muthukumar, M. (1999) *J. Chem. Phys.*, **111**, 10371.
3. Luo, K., Huopaniemi, I., Ala-Nissila, T., and Ying, S.-C. (2006) *J. Chem. Phys.*, **124**, 114704.
4. Luo, K., Huopaniemi, I., Ala-Nissila, T., and Ying, S.-C. (2006) *J. Chem. Phys*, **125**, 124901.
5. Panja, D. and Barkema, G. (2008) *Biophys. J.* **94**, 1630.
6. Gauthier, M.G. and Slater, G.W. (2009) *Phys. Rev. E*, **79**, 021802.
7. Kantor, Y. and Kardar, M. (2004) *Phys. Rev. E*, **69**, 021806.
8. Chuang, J., Kantor, Y., and Kardar, M. (2001) *Phys. Rev. E*, **65**, 011802.
9. Grosberg, A.Y., Nechaev, S., Tamm, M., and Vasilyev, O. (2006). *Phys. Rev. Lett*, **96**, 228105.
10. Vanderzande, C. (1998) *Lattice Models of Polymers*, Cambridge University Press, Cambridge.
11. Metzler, R. and Klafter, J. (2000) *Physics Rep.*, **339**, 1.
12. Metzler, R. and Klafter, J. (2003) *Biophys. J.*, **85**, 2776.
13. Wiggin, M., Tropini, C., Tabard-Cossa, V., Jetha, N.N., and Marziali, A. (2008) *Biophys. J.*, **95**, 5317–5323.
14. Oldham, K. and Spanier, J. (1974) *The Fractional Calculus: Theory and Applications of Differentiation and Integration to Arbitrary Order*, Academic Press.

15 Risken, H. (1989) *The Fokker–Planck Equation*, Springer Verlag, Berlin.
16 Milchev, A., Binder, K., and Bhattacharya, A. (2004) *J. Chem. Phys.*, **121**, 6042.
17 Gorenflo, R., Loutchko, J., and Luchko, Yu (2002). *Fract. Calc. Appl. Anal.*, **5**, 491.
18 Dubbeldam, J.L.A., Milchev, A., Rostiashvili, V.G., and Vilgis, T.A. (2007) *Phys Rev. E*, **76**, 010801.
19 Dubbeldam, J.L.A., Milchev, A., Rostiashvili, V.G., and Vilgis, T.A. (2007) *Europhys. Lett.*, **79**, 18002.
20 Lubensky, D.K. and Nelson, D.R. (1999) *Biophys. J.*, **77**, 1824.
21 Zwolak, M. and Di Ventra, M. (2008) *Rev. Mod. Phys.*, **80**, 141.

Part Two
Fundamental Aspects

7
Entropy Production, the Breaking of Detailed Balance, and the Arrow of Time
Christian Van den Broeck

7.1
Introduction

The arrow of time has preoccupied man since the dawn of civilization. It is arguably the strongest constraint to which we are exposed. Time however appears in a very different way in the theories of Newton and Einstein and in quantum mechanics. Especially the fact that the laws of physics are time-reversible and time-translation invariant seems to contradict the unidirectional flow of time. The second law of thermodynamics on the other hand stands out by its explicit statement that the entropy of a total isolated system always evolves in one direction, namely, toward a maximum. The reconciliation of the reversible time of physics with the irreversible time, which rules so much of the world around and makes time into history, has remained a puzzle. Is the irreversible time a result of the expansion of the universe or the collapse of the wavefunction? Is it just a statistical law, expressing the tautological observation that a system is more likely to evolve from a less probable to a more probable state? Does chaos play a role? What about quantum chaos (if it exists at all)? Does the arrow of time apply to nonextensive systems? Is the second law limited to the macroscopic realm? Does it reflect our imperfection of measurement, control, or manipulation? What, if any, is the microscopic expression of the entropy? Is it defined for a system out of equilibrium? Is there a microscopic proof for the increase of entropy? These issues are particularly intriguing since, according to Einstein, thermodynamics is the only physical theory of general nature that will never be overthrown. It places rigorous limits on how much work can be derived from heat baths. It has played a central role in, or is consistent with, momentous discoveries in physics, including the quantum nature of blackbody radiation, the indistinguishability of particles, and the phenomenon of stimulated emission.

The recent discoveries of the fluctuation and work theorems [1–12] have reinvigorated the debate. With respect to the average value of work or entropy production, they are consistent with the second law, but provide no additional information beyond it. Inspired by these developments, the exact value of the (average) entropy production, upon bringing a system from one canonical equilibrium state into another, has

The Complexity of Dynamical Systems. Edited by J. Dubbeldam, K. Green, and D. Lenstra
Copyright © 2011 WILEY-VCH Verlag GmbH & Co. KGaA, Weinheim
ISBN: 978-3-527-40931-0

recently been derived from first principles [13] (see also Ref. [14]). The purpose of this chapter is to make this result accessible to a wider audience, through a more informal discussion, while also making the connection with the so-called stochastic thermodynamics.

We stress from the onset that we will not resolve the basic issue of reversibility versus irreversibility. Rather we will work in the footsteps of Onsager: by confronting the assumed irreversibility with the reversibility and time translation invariance of the underlying laws, he obtained fundamental results such as the symmetry of Onsager coefficients and the concept of detailed balance, discoveries that in turn led to other important results such as the fluctuation dissipation theorem and the Green–Kubo expressions for transport coefficients.

The outline of the paper is as follows. We begin by reviewing the concept of detailed balance. Next, we apply it to stochastic thermodynamics and show that the entropy production is essentially the breaking of detailed balance. We derive the entropy production from microscopic laws and confirm again that entropy production is tantamount to the presence of an arrow of time. Finally, we close with a critical discussion.

7.2
Detailed Balance

According to the second law, no work can be extracted from a single heat bath at equilibrium. From a microscopic point of view, the more profound statement is that the processes occurring in a system at equilibrium must obey detailed balance: every process must be as likely as its time reverse. In particular, at the macroscopic level, there can be no fluxes whatsoever. Otherwise, a device sensitive to this current could be exploited to extract work, like in a windmill or a ratchet. We conclude that there is, in the macroscopic state of a system at equilibrium, no change in time. Time is, so to say, absent. It is however more revealing to illustrate the concept of detailed balance to nonextensive subparts of the equilibrium system. If the system is not at zero temperature, the observation of such smaller parts will reveal constant change and motion. Detailed balance imposes a very stringent condition on these fluctuations, since each such fluctuation and its time reverse must be equally probable. In other words, it is impossible to decide, from the statistical observation of these fluctuations, whether time is running forward or backward. There is, statistically speaking, no arrow of time. A windmill-like argument linking detailed balance to the second law has been the object of a long-standing controversy. In fact, very soon after the formulation of the second law, Maxwell introduced his famous Maxwell demon, suggesting that the increase of entropy was limited to the macroscopic realm. Many alternative constructions have been proposed that seem to violate the second law by giving the impression that one can rectify equilibrium fluctuations. This erroneous impression is often the result of applying macroscopic intuition to small-scale asymmetric objects. The fact of the matter is that detailed balance rules out the extraction of energy via, for example, small-scale ratchet-like objects. Also for such devices, fluctuating motion in one way or the other will be perfectly time symmetric in a system in equilibrium. In summary, we conclude that the basic feature of a system at equilibrium, detailed balance, corresponds not to the absence of time but to the absence of a time directionality or arrow of time.

We briefly sketch the proof of detailed balance for a classical isolated system (see Figure 7.1). The statistical state of such a system at equilibrium is described by the microcanonical ensemble. For simplicity, we assume that the total energy is specified. Every microstate (point in phase space) at that energy is equally probable. Consider now two coarse-grained regions of phase space, Ω and Ω', each corresponding to a set of microstates of nonvanishing measure and being even in the momenta. The probability to observe a transition in the state of the system, going from Ω at time zero into the state Ω' at a time between t and $t + dt$, is obviously equal to the measure of the set of microstates that make this transition. But for every such single microscopic trajectory, there is exactly one trajectory, namely, the one with reversed momenta, which will make the inverse transition. By invoking Liouville's theorem, namely, that the measure of a region in phase space is invariant under Hamiltonian dynamics, we conclude that the measure of the sets corresponding to forward and backward transitions between Ω and Ω' are equal. Hence, the probability for observing these

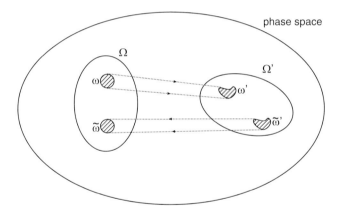

Figure 7.1 Illustration of the principle of detailed balance in a system at microcanonical equilibrium. All the accessible microstates in phase space are equally probable. Consider two states of the system, corresponding to all the microstates in the regions Ω and Ω', respectively. Let the dashed region ω, subset of Ω, represent all the microstates that make a transition from Ω to Ω' in a specific time t. In other words, the set of microstates ω evolves under Hamiltonian evolution during a time t into ω', subset of Ω'. While the shape of this set of microstates will change, its volume is preserved under Hamiltonian dynamics according to Liouville's theorem. Hence, the volume of ω is equal to that of ω'. Since all states in the accessible phase space are equally probable, the probability to observe this transition is the volume of ω divided by the total accessible space volume. To calculate the probability for the inverse transition, we note that the only trajectories that will make the inverse transition in the same amount of time are those obtained from ω', by inverting the momenta. This region $\tilde{\omega}'$ has the same volume as ω', and hence as ω. The probability to observe such a transition, that is, to pick a microstate in $\tilde{\omega}'$, is thus equal to that for the direct transitions.

transitions is equal. The argument can be repeated when one specifies additional coarse-grained states of the system at intermediate times. We conclude that the observation of any kind of process (of nonzero measure) during any time interval is as probable as its time reverse.

7.3
Stochastic Thermodynamics

Detailed balance will be broken when the system is brought out of equilibrium. The question immediately arises whether there is any connection with the entropy production. Such a connection has been established on a macroscopic level a long time ago, mainly through the contributions of L. Onsager and I. Prigogine, in the formulation of irreversible thermodynamics based on the hypothesis of local equilibrium. In this approach, it is assumed that the deviations from equilibrium are small both in time and space. More precisely, the entropy of a small region is given by its equilibrium expression in terms of the local values of the conserved variables [15, 16]. The resulting entropy production is given by the famous bilinear expression in terms of the appropriate thermodynamic fluxes J_v and forces X_v : $(d_i S/dt) = \sum_v J_v X_v$. In particular, one identifies the regime of linear irreversible thermodynamics when there is a linear relation between fluxes and forces: $J_v = \sum_{v'} L_{v,v'} X_{v'}$. The entropy production vanishes in the absence of fluxes. Onsager showed that reversibility of the microscopic laws implies the symmetry of the Onsager matrix L. Prigogine pointed out that the steady states in the linear regime are characterized by minimum entropy production.

Can one now provide a similar description at the level of small-scale systems? Such a formalism, for which we coined the name "stochastic thermodynamics" [17], has indeed been developed mostly in the context of stochastic Markovian descriptions [17–19]. It has since been considerably extended, including to what could be called the formulation of pathwise stochastic thermodynamics [6, 20–24]. We illustrate the basic procedure of stochastic thermodynamics for a Markovian master equation. We assume that the state of the system under consideration can be described by a parameter i, which for simplicity we assume to be discrete. Note that with the identification of this variable, we basically imply that any other variables (not only the internal but also external degrees of freedom, associated with the environment or medium) are irrelevant for the further description. They are, to make the analogy with the macroscopic theory, fast variables whose statistical properties, on the timescale of processes involving the relevant coordinate i, are always at equilibrium. In particular, their contribution to the entropy does not change in time, and the relevant entropy of the system is given by $S(t) = -\sum_i p_i(t) \ln p_i(t)$ (Boltzmann's constant $k_B = 1$). Here $p_i(t)$ is the probability of finding the system in state i at time t. This ansatz for the entropy is to be compared with the local equilibrium expression for the entropy used in macroscopic irreversible thermodynamics. The overbar suggests that one can also consider the quantity $\overline{S}(t) = -\ln p_i(t)$ as the instantaneous entropy of the system

7.3 Stochastic Thermodynamics

while in state i at time t. Note that there is no ambiguity with the above definitions of entropy when considering discrete states. For continuous variables, p becomes a probability density, and one runs into the familiar problem that the entropy is not invariant upon a change of variables. This problem can be solved by considering only entropy differences, by working with relative entropy (see below), or by invoking quantum mechanics.

The elimination of degrees of freedom usually implies for a small system that its dynamics becomes stochastic. One of the simpler and often quite realistic assumptions is to assume Markovianity in the variable i, resulting in a dynamics described by the master equation:

$$\dot{p}_i(t) = \Sigma_j W_{ij} p_j(t). \tag{7.1}$$

A few comments about the so-called transition matrix are in place. W_{ij} has, for $i \neq j$, the meaning of transition rate per unit time from state j to state i. Furthermore, one has that $\Sigma_i W_{ij} = 0$, or $W_{jj} = -\Sigma_{i, i \neq j} W_{ij}$. This property ensures that normalization is preserved under the dynamics. It allows to write the master equation under the balance form $\dot{p}_i(t) = \Sigma_{j, j \neq i}[W_{ij} p_j(t) - W_{ji} p_i(t)]$ making more explicit the increase and decrease of probability in the state i due to incoming and outgoing transitions, respectively. To go beyond the pure mathematical formulation, it will be essential to identify the different physical processes ν, taking place in the system. In particular, a specific transition, say from state j to state i, may occur via different mechanisms ν, which need to be specified. In the simplest case, these mechanisms do not interfere, and one can write: $W_{ij} = \Sigma_\nu W_{ij}^{(\nu)}$.

We now identify a positive quantity, which we will refer to as the rate of entropy production. Combining the master equation (7.1) with conservation of probability ($\Sigma_i \dot{p}_i = 0$, $\Sigma_{i,\nu} W_{ij}^{(\nu)} = 0$), one can write the following:

$$\dot{S} = -\sum_i p_i \frac{1}{p_i} \dot{p}_i - \sum_i \dot{p}_i \ln p_i, \tag{7.2}$$

$$= -\sum_{i,j,\nu} W_{ij}^{(\nu)} p_j \ln p_i, \tag{7.3}$$

$$= \sum_{i,j,\nu} W_{ij}^{(\nu)} p_j \left\{ \ln \frac{W_{ij}^{(\nu)} p_j}{W_{ji}^{(\nu)} p_i} + \ln \frac{W_{ji}^{(\nu)}}{W_{ij}^{(\nu)}} \right\}. \tag{7.4}$$

Hence, the rate of change of the system entropy can be written in the form of a balance equation, namely, $\dot{S}(t) = \dot{S}_i(t) + \dot{S}_e(t)$. $\dot{S}_e(t)$ is the entropy flow into the system:

$$\dot{S}_e(t) = \sum_{i,j,\nu} W_{ij}^{(\nu)} p_j \ln \frac{W_{ji}^{(\nu)}}{W_{ij}^{(\nu)}}. \tag{7.5}$$

$\dot{S}_i(t)$ is the nonnegative total entropy production for the physical processes represented by the master equation, namely,

$$\dot{S}_i = \sum_{i,j,\nu} W_{ij}^{(\nu)} p_j \ln \frac{W_{ij}^{(\nu)} p_j}{W_{ji}^{(\nu)} p_i} = \frac{1}{2} \sum_{i,j,\nu} \left\{ W_{ij}^{(\nu)} p_j - W_{ji}^{(\nu)} p_i \right\} \ln \frac{W_{ij}^{(\nu)} p_j}{W_{ji}^{(\nu)} p_i} \geq 0. \qquad (7.6)$$

For comparison with the following, we make a number of additional comments concerning the above expression for the entropy production. First, we show that the above introduced entropy production and entropy flow agree with the usual thermodynamic definitions for an explicit physical model describing the exchange of particles and energy with a number of reservoirs ν [25]. We assume that these reservoirs are and remain at equilibrium. Otherwise, we would need to include a description of their evolution as well. The interesting case arises when the reservoirs have different inverse temperatures β_ν and chemical potentials μ_ν, respectively. The system will operate as a (small-scale) nonequilibrium device, trying to restore the equilibrium between the reservoirs. This will be a "hopeless task" in the assumed limit that the reservoirs remain unaffected by their exchanges with the system: we have the perfect ingredients for the appearance of a stochastic nonequilibrium steady state in the system. At the steady state, while the overall net fluxes of particles and energy to the system will be zero, the separate fluxes to each of the reservoirs will not vanish, and a concomitant irreversibility and entropy production will be identified through the breaking of detailed balance in those separate processes. One of the most common and expedient assumptions concerning the separate rate matrices $W_{ij}^{(\nu)}$ is to assume, as said before, that the coupling of the system to the different reservoirs does not influence the dynamics of exchange with each of them. As a result, the separate rate matrices $W_{ij}^{(\nu)}$ are identical to those when the system is in contact with each of the reservoirs separately. To reproduce the grand canonical distribution, which has to prevail when the system is at the steady state (hence equilibrium) in contact with a single reservoir ν, $W_{ij} = W_{ij}^{(\nu)}$, the transition rates need to satisfy detailed balance with respect to this distribution at the prevailing temperature and chemical potential (ε_i and N_i being the energy and number of particles in state i, respectively):

$$\frac{W_{ji}^{(\nu)}}{W_{ij}^{(\nu)}} = \exp\{\beta_\nu [(\varepsilon_i - \varepsilon_j) - \mu_\nu (N_i - N_j)]\}. \qquad (7.7)$$

It is a revealing exercise to show that indeed for the specific choice of transition rates mentioned before, the steady-state entropy flow and entropy production are given by the familiar thermodynamic expressions in terms of the thermodynamics fluxes and forces [25] (see also Refs [17–19] for other examples). For example, inserting (7.7) into the expression for the entropy flux, we find

$$\dot{S}_e(t) = \sum_\nu \beta_\nu \sum_{i,j} W_{ij}^{(\nu)} p_j [(\varepsilon_i - \varepsilon_j) - \mu_\nu (N_i - N_j)]. \qquad (7.8)$$

We recognize the exchange of internal energy $\varepsilon_i - \varepsilon_j$ and chemical work $\mu_\nu (N_i - N_j)$ exchanged with bath ν upon transition from j to i. Multiplied with the frequency $W_{ij}^{(\nu)} p_j$ of these transitions, one finds that $\dot{S}_e(t)$ is equal to the sum over all reservoirs, of the rate of heat transfer (being internal energy transfer minus chemical work transfer) from each of them into the system, divided by the temperature. In a similar way, one recovers the usual expression for the entropy production [25].

As a second comment, it is clear that the entropy production is determined by the breaking of detailed balance, that is, by the difference in frequency between $W_{ij}^{(\nu)} p_j$ and $W_{ji}^{(\nu)} p_i$, corresponding to the i to j versus j to i transitions, via the mechanism ν. Here, we note again the crucial importance of identifying the separate physical processes. Failure to do so will in fact lead to an underestimation of the entropy production. This important observation is missing in the first formulation of the theory [18], and was pointed out subsequently in Ref. [17]. The underestimation follows from the following inequality:

$$\sum_{\nu} W_{ij}^{(\nu)} p_j \ln \frac{W_{ij}^{(\nu)} p_j}{W_{ji}^{(\nu)} p_i} \geq W_{ij} p_j \ln \frac{W_{ij} p_j}{W_{ji} p_i} \quad \forall (i,j), \tag{7.9}$$

implying

$$\dot{S}_i = \sum_{i,j,\nu} W_{ij}^{(\nu)} p_j \ln \frac{W_{ij}^{(\nu)} p_j}{W_{ji}^{(\nu)} p_i} \geq \sum_{i,j} W_{ij} p_j \ln \frac{W_{ij} p_j}{W_{ji} p_i} = \dot{S}_i^c. \tag{7.10}$$

We introduced here the coarse-grained entropy $\dot{S}_i^c$, which is always smaller than the true full entropy production. The above result is easily proven from the so-called log-sum inequality [26], valid for nonnegative numbers a_ν and b_ν:

$$\sum_{\nu} a_\nu \ln \frac{a_\nu}{b_\nu} \geq \sum_{\nu} a_\nu \ln \frac{\sum_\nu a_\nu}{\sum_\nu b_\nu}, \tag{7.11}$$

by considering the choice $a_\nu = W_{ij}^{(\nu)} p_j$ and $b_\nu = W_{ji}^{(\nu)} p_i$ for $i \neq j$.

The above subtlety in identifying the basic processes is further illustrated by the following extreme case. We consider the chemical reaction $A \underset{}{\overset{k_1}{\rightarrow}} X \underset{}{\overset{k_2}{\rightarrow}} B$, where A and B are reservoir variables kept at fixed concentrations. The stochastic evolution in the number of X particles (for simplicity of notation, we represent by A, B, and X the number of particles of the respective species) is typically described by a Markovian master equation [27]. We will furthermore assume that the above reactions are far out of equilibrium, with backward transitions taking place extremely rarely. It is then, for mathematical convenience, tempting to assume that the reactions are in fact completely irreversible. The transition rates for the corresponding Markovian master equation are $W_{X+1,X}^1 = k_1 A$ and $W_{X,X+1}^2 = k_2 (X+1)$, with k_1 and k_2 representing the rate constants for the first and second chemical reaction, respectively. The backward transition rates are taken to be zero, $W_{X,X+1}^1 = 0$ and $W_{X+1,X}^2 = 0$. We conclude from the formula (7.6) that the rate of entropy production diverges. If however, one uses the coarse-grained transition rates $W_{X+1,X} = k_1 A$ and $W_{X,X+1} = k_2(X+1)$, one finds that, at the steady state, a coarse-grained version of "detailed balance" will hold, $W_{X+1,X} p(X) = W_{X,X+1} p(X+1)$, and the coarse-grained entropy production is zero! The procedure to replace highly irreversible processes by completely irreversible ones is often used, since it usually simplifies the mathematical description of the system. From the point of view of entropy production however, it is a singular limit.

Third, the positivity of (7.6) can be viewed as a purely mathematical property: we do not need to make any physical assumptions on the transition probabilities. The agreement with thermodynamics with an appropriate choice of the transition probabilities could be fortuitous; hence, a more microscopic argument would be welcome. To make the connection with such a theory presented below, we now derive an expression for the integrated rate of entropy production or the entropy difference by considering paths of the stochastic process. For simplicity, we restrict ourselves to the case of steady states. We are interested in the probability for seeing a specific succession of states, namely, $(i_1; i_2; i_3; \ldots; i_n)$, at (closely, equally spaced) discrete times $t_k = t_1 + (k-1)dt$, $k = 1, \ldots, n$, versus that of the reverse sequence $(i_n; i_{n-1}; \ldots; i_1)$. By invoking Markovianity, and the fact that for a small time difference the conditional probability is proportional to the transition rate, $P(i_k|i_{k-1}) = W_{i_k, i_{k-1}} dt$ (for $i_k \neq i_{k-1}$, note that the terms $i_k = i_{k-1}$ drop out), one now has for the log ratio of probabilities of these events

$$\ln \frac{P(i_1; i_2; i_3; \ldots; i_n)}{P(i_n; i_{n-1}; \ldots; i_1)} = \ln \frac{p_{i_1} \Pi_{k=2,n} W_{i_k i_{k-1}}}{p_{i_n} \Pi_{k=2,n} W_{i_{k-1} i_k}} = \sum_{k=2,n} \ln \frac{W_{i_k i_{k-1}} p_{i_{k-1}}}{W_{i_{k-1} i_k} p_{i_k}}. \quad (7.12)$$

By averaging this quantity with the probability distribution for the trajectory, we conclude that

$$D(P(i_1; i_2; \ldots; i_n) \| P(i_n; i_{n-1}; \ldots; i_1)) = \Sigma_{i_1; i_2; \ldots; i_n} P(i_1; i_2; \ldots; i_n) \ln \frac{P(i_1; i_2; \ldots; i_n)}{P(i_n; i_{n-1}; \ldots; i_1)}$$

$$= \Sigma_{k=2,n} dt \Sigma_{i_k, i_{k-1}} W_{i_k i_{k-1}} p_{i_{k-1}} \ln \frac{W_{i_k i_{k-1}} p_{i_{k-1}}}{W_{i_{k-1} i_k} p_{i_k}} = \int dt \dot{\bar{S}}_i^c = \Delta \bar{S}_i^c, \quad (7.13)$$

which is the requested integrated version of the entropy production. Note that it gives the coarse-grained entropy difference, since by using only the states of the Markov process, we do not distinguish between different mechanisms of transitions between the states. We have also introduced another object, which will play a central role in the sequel, namely, the relative entropy $D(p\|q) = \langle \ln p/q \rangle$ (with average with respect to p) of two probability distributions p and q. We conclude that the (coarse-grained) entropy production is the relative entropy of the probabilities for the paths and their time-reversed counterparts. The relative entropy is a measure of a (asymmetric) distance between these distributions, and, for example, linked to the difficulty of distinguishing samplings from each of them [26]. Hence, the above expression relates the entropy production to the arrow of time.

Fourth, we want to make the connection with trajectory-dependent quantities. Coming back to the expression for the total entropy production $\bar{S}_i(t)$, we note that it can be considered as the average of the following quantity:

$$\dot{S}_i(t) = \ln \frac{W_{ij}^{(v)} p_j}{W_{ji}^{(v)} p_i}. \quad (7.14)$$

In the same line of thought, we can consider

$$\Delta S = \ln \frac{P(i_1; i_2; i_3; \ldots; i_n)}{P(i_n; i_{n-1}; \ldots; i_1)} \tag{7.15}$$

as the (coarse-grained) trajectory-dependent entropy production for the sequence $(i_1; i_2; i_3; \ldots; i_n)$ (we dropped the superscript c for simplicity of notation here). The interest in this quantity is further enhanced by noting that its probability distribution obeys a symmetry relation, known as the fluctuation theorem, namely,

$$\frac{P(\Delta S)}{P(-\Delta S)} = e^{\Delta S}. \tag{7.16}$$

In other words, the probability to observe a (positive) value ΔS is exponentially more likely than that of observing $-\Delta S$. It is important to stress, once more, that from a mathematical point of view, the above relation is in fact an (almost trivial) identity that requires no physical assumptions. In fact, it is valid independent of any assumption on the dynamics generating the probability for the trajectory. The following proof highlights the origin of the relation (7.16). Consider $x \in \Omega$, a random variable (possibly in a multidimensional space, for example, $x = \{i_1; i_2; i_3; \ldots; i_n\}$) with probability distribution $P(x)$. Let T be an involution on the space Ω, that is, $T(x) \in \Omega, \forall x \in \Omega$, with $T(y) = T(T(x)) = x$. The Jacobian of such a transformation from x to $y = T(x)$ has necessarily an absolute value equal to 1. Note that the above considered time reversal $T(x) = T(\{i_1; i_2; i_3; \ldots; i_n\}) = \{i_n; i_{n-1}; \ldots; i_1\}$ is a simple example of an involution. We now introduce the following new random variable:

$$\sigma(x) = \ln \frac{P(x)}{P(T(x))}. \tag{7.17}$$

This quantity measures the difference in surprise of observing the outcome $T(x)$ as opposed to x. We can write the following set of equalities for the probability distribution $P(\sigma)$ for the quantity σ:

$$P(\sigma) = \int dx P(x) \delta\left(\ln \frac{P(x)}{P(T(x))} - \sigma\right) = \int dx e^\sigma P(T(x)) \delta\left(\ln \frac{P(x)}{P(T(x))} - \sigma\right)$$

$$= e^\sigma \int dy P(y) \delta\left(\ln \frac{P(T(y))}{P(y)} - \sigma\right) = e^\sigma P(-\sigma), \tag{7.18}$$

where in transition from line 1 to line 2, we used the delta function that allows to replace $P(x)$ by $\exp(\sigma)P(T(x))$, and in transition from line 2 to line 3, we used the Jacobian of the transformation for x to $y = T(x)$ (with $x = T(y)$) is equal to 1. We conclude that $P(\sigma)$ obeys the fluctuation theorem. This mathematical property is of special interest in the above context, since the quantity ΔS, directly related to the time irreversibility of a specified trajectory, has, under the appropriate physical conditions, the important meaning of the (coarse-grained) entropy production along this trajectory.

It is also revealing and useful to note that one can separate the entropy production in a "medium" and "system" contribution. Indeed, one can write $\Delta S = \Delta S_m + \Delta S_s$, with

$$\Delta S_m = \ln \frac{P(i_2; i_3; \ldots; i_n | i_1)}{P(i_{n-1}; \ldots; i_1 | i_n)} \tag{7.19}$$

playing the role of the entropy production in the medium and $\Delta S_s = \ln(P(i_1)/P(i_n))$ being the change in entropy of the system. In terms of the medium entropy production, one can derive a so-called integral fluctuation theorem [21], namely,

$$\left\langle e^{-\Delta S_m} \frac{p'_{i_n}}{p_{i_1}} \right\rangle = 1, \tag{7.20}$$

which is valid for any choice of the probability distribution p'. The average is performed with respect to the forward probability distribution $P(i_1, i_2; i_3; \ldots; i_n)$, and the result follows from normalization. For the specific choice of $p' = p$, one recovers the "usual" integral fluctuation theorem $\langle \exp(-\Delta S) \rangle = 1$, which is also a direct consequence of the fluctuation theorem itself.

The system entropy, being the difference between the "full" entropy and the "medium" entropy contribution, appears as a "boundary term." One would naively expect that upon measuring the trajectory-wise entropy along an asymptotically long trajectory, this boundary term becomes irrelevant, so that the fluctuation theorem is also valid for the medium entropy. However, the fluctuation theorem deals with large deviations, and in case of a system with unbounded energy (or state space), exponentially rare initial conditions may dominate the tail of the distribution of the total entropy production. This observation was first made by Farago [28], and further elaborated in Ref. [29].

To conclude this discussion of trajectory-dependent quantities, we finally note that all the above considerations can be generalized to time-dependent transition probabilities and nonsteady-state conditions as well as to other Markovian descriptions, such as the one based on Langevin or Fokker–Planck equations. This formalism thus provides a very elegant and detailed stochastic thermodynamic description of such nonequilibrium systems (see, for example, Refs [22–25]). Nevertheless, a complementary microscopic theory would clearly be welcome. Such a theory is the object of the next section.

7.4
Microscopic Expression of Entropy Production

We will now derive an exact microscopic expression for the entropy production in the context of a simple scenario. It turns out that the final result is much more general than its derivation suggests, and we will also give a handwaving argument for its robustness. The scenario is the following. The system is described by the Hamiltonian $H(q, p; \lambda)$, where q, p denote all microscopic degrees of freedom (position and momenta) and λ is a control parameter that describes the energy exchange with an external mechanical device. For simplicity, we will assume that the Hamiltonian is an even function with respect to inversion of momenta. The system is initially assumed to be in canonical equilibrium at temperature T at the value λ_A of the control parameter. This can be achieved by putting it in contact with a heat bath at this temperature, which is subsequently disconnected. Next, the control parameter is changed following a specific protocol $\lambda(t)$ from λ_A to λ_B. In the process, we will assume that no energy is exchanged other than the work W performed by the external agent on the system. We will furthermore consider the time-reversed scenario, with the system initially at canonical equilibrium at the same temperature T, but at the value λ_B of the control parameter. The latter is now changed according to the exact time-reversed protocol. We will indicate with a superscript tilde all the variables that relate to this so-called reverse experiment.

We first set out to calculate the work $W(q, p; t)$ done along the forward process for a specific phase trajectory, namely, the one that passes through the phase point (q, p) at time t. Since the dynamics are deterministic, there is precisely one such trajectory. Let us call (q_0, p_0) and (q_1, p_1) the corresponding initial and final phase points uniquely determined by (q, p) (for the specified scenario $\lambda(t)$). Also note that there is a one-to-one correspondence with the time-reversed trajectory in the time-reversed protocol, which starting from $(q_1, -p_1)$ goes through $(q, -p)$ and, finally, to $(q_0, -p_0)$. For simplicity of notation, we will use the forward time to express times in both forward and backward scenarios. By conservation of total energy, one has that $W(q, p; t) = H(q_1, p_1; \lambda_B) - H(q_0, p_0; \lambda_A)$. Now, since the phase-space density is conserved along any Hamiltonian trajectory, one has in both forward and backward process,

$$\varrho(q, p; t) = \varrho(q_0, p_0; t_0) = \frac{\exp[H(q_0, p_0; \lambda_A)]}{Z_A} \quad (7.21)$$

$$\tilde{\varrho}(q, -p; t) = \tilde{\varrho}(q_1, -p_1; t_1) = \frac{\exp[H(q_1, -p_1; \lambda_B)]}{Z_B}, \quad (7.22)$$

where Z_A and Z_B are partition functions at the equilibrium states A and B, respectively. These expressions allow us to eliminate the Hamiltonian (which is supposed to be even in the momenta) at initial and final times in favor of the phase-space density at any intermediate time point:

$$W(q, p, t) - \Delta F = T \ln \frac{\varrho(q, p, t)}{\tilde{\varrho}(q, -p, t)} = \ln \frac{\text{Prob}(\text{path}(q, p, t))}{\tilde{\text{Prob}}(\tilde{\text{path}}(q, -p, t)}. \quad (7.23)$$

$\Delta F = kT \ln Z_B/Z_A$ is the free energy difference between the final and initial equilibrium states. Note that the (phase-space) probability density $\varrho(q, p, t)$ and $\tilde{\varrho}(q, -p, t)$ for being in the microstates (q, p, t) and $(q, -p, t)$ at the forward time in the forward and reverse experiments are equal to the probability density for observing the corresponding paths, which we represented by $\text{Prob}(\text{path}(q, p, t))$ and $\tilde{\text{Prob}}(\tilde{\text{path}}(q, -p, t)$. The average dissipated work is obtained by noting that $W(q, p, t)$ will appear with the probability density $\varrho(q, p, t)$, hence,

$$\langle W(q,p,t)\rangle - \Delta F = TD(\varrho(q,p,t)||\tilde{\varrho}(q,-p,t)). \tag{7.24}$$

This equation features the dissipated work, defined as the extra amount of work, on top of the difference of free energy ΔF, which is required for making the transition.

To make the connection with the entropy production (in the forward process), a further discussion is required. Note that at the end of the forward process, the system is generally speaking in a genuine nonequilibrium state for which the entropy is not well defined. By genuine, we mean that one cannot, in general, make an assumption of local equilibrium. This assumption, invoking an equilibrium-like expression of the entropy in terms of a reduced number of variables (the other deemed to be at equilibrium) is, as said before, frequently and successfully invoked at the level of macroscopic thermodynamics and lies also at the basis of stochastic thermodynamics. Defining the entropy production on the microscopic scene is however much more delicate and remains an issue to be solved. In particular, it is known that $-\int dq\, dp\, \varrho(q,p,t)\ln \varrho(q,p,t)$ is preserved under Hamiltonian evolution. Hence, it does not seem appropriate to use this expression for the entropy of a nonequilibrium state (see also some suggestions on this issue Section 7.5). To circumvent this problem, we introduce an idealized heat bath at temperature T, with which the system is put in contact at the end of the forward experiment, without extra work. The system thus relaxes back to canonical equilibrium, which is also the starting state of the backward process. During the relaxation, the system absorbs an average amount $\langle Q \rangle$ of heat for the bath. Note that we are referring here to an average for two reasons. First, the system and bath will continuously keep exchanging energy, back and forth, due to the presence of fluctuations, and second, there is also an average over the trajectories (or canonical initial conditions) of the system. We now evaluate the total (average) entropy change of system plus heat bath. On the one hand, the entropy change in the bath, assuming it operates quasi-statically, is given by $\Delta \bar{S}_m = -\langle Q \rangle/T$. On the other hand, the entropy change between the canonical equilibrium states of initial and final states in the system reads $\Delta \bar{S}_s = (\Delta U - \Delta F)/T = (\langle Q \rangle + \langle W \rangle - \Delta F)/T$. Here, we used the equilibrium relation $F = U - T\bar{S}_s$, $U = \langle H \rangle$ is the equilibrium internal energy, calculated with respect to the prevailing canonical distribution, and $\bar{S}_s$ is the thermodynamic entropy of the system (the bar was introduced to distinguish it from the trajectory-dependent quantity). The total entropy production (in the total device, system plus heat bath) in the forward process, $\Delta \bar{S}$, is thus equal to the dissipated $\langle W \rangle - \Delta F$ divided by temperature; hence,

$$\Delta \bar{S} = D(\varrho || \tilde{\varrho}) = D(\text{Prob}(\text{path}) || \tilde{\text{Prob}}(\tilde{\text{path}})). \tag{7.25}$$

Written in this form, the similarity with the steady-state coarse-grained entropy from (7.13) is obvious. Note that the formula (7.13) applies to steady-state conditions, explaining why the probability of the backward path in (7.13) can be evaluated with the same distribution. Note also that we obtain here the full and not the coarse-grained entropy like in (7.13), because on a microscopic level, there is no degeneracy in the transition mechanism when the system changes its state. We also stress again the essential role, in the above derivation, of Liouville's theorem guaranteeing that a phase-space volume is preserved under canonical transformation, particularly under Hamiltonian evolution. Hence, the Jacobian of any canonical transformation is equal to 1, and the above quantity is independent of the choice of canonical variables.

We claim that (7.25) is a fundamental expression for the entropy production. One can indeed show that the same expression is obtained in many other scenarios of transitions between equilibrium states, and also applies for a quantum mechanical description (with ϱ being the density matrix) [30]. Here, we limit ourselves to a handwaving argument, which at the same time identifies the microvariables (q, p) that need to be taken into account for the application of (7.25). We start with a large system at equilibrium. We perform a perturbation, for example, a time-dependent change of an external field, or the removal of a constraint that allows the flow of energy and particles, but assume that this perturbation is acting on a subpart of the system. Consider now a subsystem that encompasses this subpart, but being still much smaller than the total system. This subsystem starts at canonical equilibrium (at the temperature of the entire system). Furthermore, one can assume that during the time of application, the perturbation does not reach the boundaries of the subsystem. This boundary remains at equilibrium during the process, and there is no net exchange of energy or particles. Hence, it does not matter for the irreversible processes taking place in the small perturbed region of the system whether we assume that the subsystem is disconnected or not during the process. We thus have the ingredients for the above derivation, in terms of the microvariables (q, p) describing the subsystem. Adding (q, p) variables of a larger part of the system will not matter since these variables are at equilibrium and hence have time symmetry.

Within a classical framework, one could point out two weaknesses in the above argument. First, extreme events could propagate so fast that they do reach the boundary. Even though such events will be extremely unlikely, it is not excluded that they play a role in the above formula (see discussion section). One could of course invoke relativity to argue that there is a finite speed of propagation, but that is not a "fair" argument in the present discussion based on classical mechanics. Second, and more troubling, the fact that the total entropy of the entire system is constant under Hamiltonian dynamics seems to be in contradiction with the increase we just calculated. The entropy increase is obtained in our derivation because we assume that the entire system relaxes back to equilibrium, with the larger part of the system playing the role of an ideal heat bath. This issue remains unresolved, but we will come back to it in Section 7.5.

To conclude this section, we note that the quantity ΔS

$$\Delta S = \frac{W(q, p, t) - \Delta F}{T} = \ln \frac{\varrho(q, p, t)}{\tilde{\varrho}(q, -p, t)} = \ln \frac{\text{Prob}(\text{path}(q, p, t))}{\widetilde{\text{Prob}}(\tilde{\text{path}}(q, -p, t))} \quad (7.26)$$

can be interpreted as the entropy production along the specified forward trajectory. As shown above, its average is indeed the "bona fide" entropy production, and its expression agrees with the one proposed in the stochastic description (cf. Equation 7.15).

7.5
Discussion

We conclude with a number of critical remarks. First, we reiterate that the above derivation is not entirely satisfactory. The entropy production in the heat bath and ensuing relaxation of the system are not obtained from a microscopic analysis. However, our formula captures, in the spirit of Onsager, the basic properties of Hamiltonian evolution, namely, microreversibility and incompressibility in phase space. Furthermore, this expression for entropy production can be applied, besides the Hamiltonian context or the stochastic context discussed above, in several other fields of research, including natural systems, dynamical systems, nonequilibrium steady states described by SRB measures, and thermostated systems [3, 18, 19, 21, 31–43].

Second, we have hinted at the connection between the microscopic formula and its stochastic counterpart. While the similarity in structure of the formulas is obvious, we have not given a specific proof relating them with each other. It turns out that one can make a clean connection without specifying the detailed procedure of deriving the stochastic description from the microscopic one. The only assumption needed is the obvious one, namely, the stochastic description is correct (i.e., it is an exact consequence, in a limit that need not be specified, of the microscopic laws). We refer to Ref. [39] for the proof of this rather satisfying observation.

Third, one may wonder whether the above formula, even if it is exact and has a wide range of application, is useful and has, aside from its purely theoretical value, practical applications. Considering the formula (7.25) for the entropy production, one can make following observation. The integral is dominated by the typical realizations of the forward process, that is, by the phase-space points (q, p) or the paths passing through these points, for which $\varrho(q, p; t)$ is large. The probability for these points or their corresponding trajectories has to be compared with the probability for the time-reversed trajectory in the backward process. When such a trajectory is atypical in the backward process, that is, when it has a low probability, then there is significant entropy production. Let us turn the argument around. Entropy is an extensive quantity. Hence, in case one is not at equilibrium, the entropy is typically proportional to the number of degrees of freedom. To reproduce this result via (7.25), we conclude that the time-reversed versions of the typical forward trajectory have to be exponentially unlikely in the backward process. This is just another way of saying that entropy production is associated with the arrow of time. Playing the movie backward will look very different when the entropy is significant. The drawback is that because of the extensivity of entropy and the corresponding exponential dependence of probabilities, the observation of the backward trajectories dominating the relative entropy in (7.25) is feasible only when we are either very close to equilibrium or when we are dealing with small systems (more precisely, with systems that have only a few degrees of freedom that are not at equilibrium). The situation is, not surprisingly, similar to that for the application of the work and fluctuation theorem [14]. This however is not the end of the story: there is one specific property that makes our formula quite useful, namely, the chain rule of relative entropy. It states that the relative entropy of two probability distributions can only become smaller upon coarse graining, that is, upon considering reduced distributions. Hence,

$$D(P, \tilde{P}) \leq D(\varrho \| \tilde{\varrho}) = \Delta \bar{S} \tag{7.27}$$

for any coarse-grained distributions P and $\tilde{P}$ of ϱ and $\tilde{\varrho}$. In other words, any observed arrow of time, that is, statistical distinction between forward and backward process, gives rise to a lower bound for the entropy production. In the absence of any information on time asymmetry, one recovers the "minimal" statement incorporated in the second law, namely, the entropy cannot decrease. The above results also suggest that the choice of variables, which strongly display time asymmetry, can capture most or all the entropy production. For example, upon quenching an external potential for a Brownian particle in a heat bath, the only variable that is not at equilibrium at the instant of the quench is the position of this particle [13]. The full entropy production is reproduced by the time asymmetry of this single variable. As one measures at later times after the quench, the signature of irreversibility moves into other variables, in particular the heat bath variables, and one needs to include their arrow of time to reproduce the total entropy production. These features can be nicely illustrated in more detail in exactly solvable harmonic models [44].

Fourth, we note the very interesting application of the above formalism to the study of computation and information processing by small systems. For example, our expression for entropy production allows a microscopic derivation of the Landauer principle [13] (see also Refs [45–49]).

Finally, returning to the issue of microscopic entropy, one can still wonder why a purely microscopic derivation seems to evade us. One possible explanation (suggested very early on in the debate, cf. the concept of Poincare recurrences) is that one needs to take the thermodynamic limit to reach irreversible behavior in any finite subpart of what is then an infinite system. In fact, explicit examples of this scenario include a particle interacting with an infinite bath of harmonic oscillators or a point particle in an infinite one-dimensional gas of hard points (with the proviso that one has to assume an equilibrium distribution for the "bath" degrees of freedom. This, one could argue, is not very different from the more *ad hoc* introduction of a heat bath). Discussing the total entropy of the latter infinite system is no longer an issue. Another related option is to imagine that while the total microscopic entropy is indeed a constant, low entropic contributions naturally build up in highly correlated processes involving many particles. This order, spread over many variables in a convoluted way, becomes *de facto* unobservable and unexploitable. It is quite likely that this "evaporation" of the entropy often has "exponential" characteristics. In a large system, the low entropy can be sucked up by an exponentially large number of correlations. In systems displaying chaotic dynamics, the exponential sensitivity to initial conditions will further enhance the phenomenon. Conversely, the "simple" degrees of freedom undergo an increase of entropy and basically dominate our perception, measurement, and interaction. The price of such a view, of course, is that entropy becomes to some extent "subjective," depending on the level of description or accuracy available. If the dependence is exponential, one can argue that the order hidden in the many degrees of freedom is for all practical purposes irretrievable and the second law prevails. A related explanation, suggested early on by Gibbs, is that while the microscopic entropy $\langle \ln \varrho \rangle$ is constant (and extremely large), coarse-grained entropies $\langle \ln P d\Omega \rangle$, where P is the integral of ϱ over some phase volume $d\Omega$, are increasing in time, as the gradients in ϱ can no longer be perceived on this coarser level [50]. A third alternative is that the importance attributed to entropy, and our obsession to provide a derivation of the second law, is perhaps a misguided historical accident. After all, the quantities of interest are the observables, and a theory should focus on the derivation of their properties. Ergodic theory can be seen as belonging to this tradition: one shows that the observed long time average of a specific quantity is identical to its (equilibrium) ensemble average. As a more recent development of this point of view, we note the recent discussion on canonical typicality, in which it is proven that very simple assumptions lead, as far as the measurements of observables are concerned, to results indistinguishable from those coming from equilibrium ensembles [51–53]. In such a derivation, there is no need to introduce the entropy or to address its increase.

In conclusion, whether the arrow of time lies in the eyes of the beholder or in his heart, we still do not know. Surprisingly, the debate has been reignited and our understanding greatly improved by the theoretical description, simulation, and experimental study of small-scale systems, which are precisely those systems in which the arrow of time is much less pronounced.

References

1 Bochkov, G.N. and Kuzovlev, Y.E. (1981) *Physica A*, **106**, 443, 480.
2 Evans, D., Cohen, E.G.D., and Morris, G.P. (1993) *Phys. Rev. Lett.*, **71**, 2401.
3 Gallavotti, G. and Cohen, E.G.D. (1995) *Phys. Rev. Lett.*, **74**, 2694.
4 Jarzynski, C. (1997) *Phys. Rev. Lett.*, **78**, 2690.
5 Jarzynski, C. (1997) *Phys. Rev. E*, **56**, 5018.
6 Crooks, G.E. (1998) *J. Stat. Phys.*, **90**, 1481.
7 Crooks, G.E. (1999) *Phys. Rev. E*, **60**, 2721.
8 Derrida, B., Gaspard, P. and Van den Broeck, C.(eds) (2007) *C.R. Physique*, **8**, 483.
9 Jarzynski, C. (2008) *Eur. Phys. J.*, **B64**, 331.
10 Ritort, F. (2003) *Semin. Poincare*, **2**, 195.
11 Liphardt, J., Dumont, S., Smith, S.B., Tinoco, I., Jr., and Bustamante, C. (2002) *Science*, **296**, 1832.
12 Collin, D., Ritort, F., Jarzynski, C., Smith, S.B., Tinoco, I., Jr., and Bustamante, C. (2005) *Nature*, **437**, 231.
13 Kawai, R., Parrondo, J.M.R., and Van den Broeck, C. (2007) *Phys. Rev. Lett.*, **98**, 080602.
14 Jarzynski, C. (2006) *Phys. Rev. E*, **73**, 046105.
15 Prigogine, I. (1947) *Etude Thermodynamique des Phenomenes Irreversibles*, Desoer, Liege.
16 de Groot, S.R. and Mazur, P. (1984) *Non-Equilibrium Thermodynamics*, Dover, New York.
17 Van den Broeck, C. (1986) Stochastic thermodynamics, in *Self-Organization by Nonlinear Irreversible Processes* (eds W. Ebeling and H. Ulbricht), Springer, Berlin, pp. 57–61.
18 Schnakenberg, J. (1976) *Rev. Mod. Phys.*, **48**, 571.
19 Luo, J.L., Van den Broeck, C., and Nicolis, G. (1984) *Z. Phys.*, **B56**, 165.
20 Sekimoto, K. (1998) *Prog. Theor. Phys. Suppl.*, **130**, 17.

21 Seifert, U. (2005) *Phys. Rev. Lett.*, **95**, 040602.
22 Seifert, U. (2008) *Eur. Phys. J.*, **B64**, 423–431.
23 Andrieux, D., Gaspard, P., Ciliberto, S., Garnier, N., Joubaud, S., and Petrosyan, A. (2007) *Phys. Rev. Lett.*, **98**, 150601.
24 Esposito, M., Harbola, U., and Mukamel, S., *Rev. Mod. Phys.*, **81**, 1665.
25 Esposito, M., Lindenberg, K., and Van den Broeck, C. (2009) *Phys. Rev. Lett.*, **102**, 130602.
26 Cover, T.M. and Thomas, J.A. (2006) *Elements of Information Theory*, 2nd edn, Wiley–Interscience, Hoboken.
27 Van Kampen, N.G. (1992) *Stochastic Processes in Physics and Chemistry*, North-Holland, Amsterdam.
28 Farago, J. (2002) *J. Stat. Phys.*, **107**, 781;28. Farago, J. (2004) *Physica A*, **331**, 69.
29 van Zon, R. and Cohen, E.G.D. (2003) *Phys. Rev. Lett.*, **91**, 110601;29.Visco, P. (2006) *J. Stat. Mech.*, P06006;29.Baiesi, M., Jacobs, T., Maes, C., and Skantzos, N.S. (2006) *Phys. Rev. E*, **74**, 021111.
30 Parrondo, J.M.R., Van den Broeck, C., and Kawai, R. (2009) *New J. Phys.*, **11**, 073008.
31 Esposito, M., Lindenberg, K., and Van den Broeck, C. (2009) *Europhys. Lett.*, **85**, 60010.
32 Mackey, M.C. (1989) *Rev. Mod. Phys.*, **61**, 981.
33 Qian, H. (2001) *Phys. Rev. E*, **63**, 042103.
34 Jiang, D.Q., Qian, M., and Zhang, F.-X. (2003) *J. Math. Phys.*, **44**, 4176.
35 Maes, C. and Netocyny, K. (2003) *J. Stat. Phys.*, **110**, 269.
36 Gaspard, P. (2004) *J. Stat. Phys.*, **117**, 599.
37 Costa, M., Goldberg, A., and Peng, C.-K. (2005) *Phys. Rev. Lett.*, **95**, 198102.
38 Porporato, A., Rigby, J.R., and Daly, E. (2007) *Phys. Rev. Lett.*, **98**, 094101.
39 Gomez-Marin, A., Parrondo, J.M.R., and Van den Broeck, C. (2008) *Europhys. Lett.*, **82**, 50002.
40 Blythe, R.A. (2008) *Phys. Rev. Lett.*, **100**, 010601.
41 Evans, D.J. and Searles, D.J. (2002) *Adv. Phys.*, **51**, 1529.
42 Pomeau, Y. (1982) *J. Phys.*, **43**, 859.
43 Weiss, J.B. (2009). *Geophys. Res. Lett.*, **36** L10705.
44 Gomez-Marin, A., Parrondo, J.M.R., and Van den Broeck, C. (2008) *Phys. Rev. E*, **78**, 011107.
45 Landauer, R. (1961) *Adv. Chem. Phys.*, **5**, 183.
46 Bennett, C.H. (1973) *IBM J. Res. Dev.*, **17**, 525.
47 Hopfield, J.J. (1974) *Proc. Natl. Acad. Sci. USA*, **71**, 4135.
48 Piechocinska, B. (2000) *Phys. Rev. A*, **61**, 062314.
49 Andrieux, D. and Gaspard, P. (2008) *Proc. Natl. Acad. Sci. USA*, **105**, 9516.
50 ter Haar, D. (1954) *Elements of Statistical Mechanics*, Rinehart & Company, Inc., New York.
51 Popescu, S., Short, A.J., and Winter, A. (2006) *Nat. Phys.*, **2**, 758.
52 Goldstein, S., Lebowitz, J.L., Tumulka, R., and Zanghi, N. (2006) *Phys. Rev. Lett.*, **96**, 050403.
53 Reimann, P. (2008) *Phys. Rev. Lett.*, **101**, 190403.

8
Monodromy and Complexity of Quantum Systems
Boris Zhilinskii

8.1
Introduction

In order to see better the place of the present chapter within the very broad subject of complex dynamical systems, we need to start with a short discussion of three key notions: quantum systems, complexity, and monodromy, leading to explaining their interrelation and relevance to the general study of dynamical systems.

By quantum systems we mean here very simple objects formed by a finite number of particles, typically atoms and molecules, which exist in bound states and can be described within the standard nonrelativistic quantum mechanics. This means that the system can exist in a number of states, characterized by discrete values of certain physical quantities or equivalently by a set of quantum numbers. The physical quantities are represented in quantum theory by operators whose eigenvalues give the possible values of these quantities. The operators do not generally commute. In order to characterize the quantum state, we need to use the eigenvalues of a set of mutually commuting operators.

The finite number of degrees of freedom and even the finite number of quantum states in which we are typically interested lead to an impression that such dynamical systems have more chance to be treated as "simple dynamical systems" than as complex. Thus, the hydrogen atom, for example, is one of the simplest real quantum systems with a completely regular set of quantum states forming highly degenerate groups of levels, so-called shells, due to specific dynamic symmetry of the problem. Such shells can be independently considered as quantum dynamical systems with two degrees of freedom using effective Hamiltonians. At the same time, even a small external perturbation of a hydrogen atom by constant electric and magnetic field leads to a splitting of the degeneracy and to the appearance of a complicated pattern of eigenstates strongly depending on the values of external parameters characterizing the perturbation.

More generally, the analyzed system of quantum states can be considered as a simple one if all states can be arranged in regular patterns characterized by several good quantum numbers taking consecutive integer values. Such quantum systems

---

The Complexity of Dynamical Systems. Edited by J. Dubbeldam, K. Green, and D. Lenstra
Copyright © 2011 WILEY-VCH Verlag GmbH & Co. KGaA, Weinheim
ISBN: 978-3-527-40931-0

correspond to classical Hamiltonian integrable systems whose dynamics is regular and is associated with toric fibrations. For an extremely simple case, one can assume that classical action–angle variables are globally defined. For quantum systems, this means that the set of quantum states represented through the joint spectrum of quantum operators can be globally arranged in a regular lattice whose nodes are characterized by quantum numbers associated with eigenstates of mutually commuting quantum operators corresponding to classical integrals of motion.

Another natural limiting behavior of dynamic systems is associated with the irregular spectra of quantum systems or with the irregular or chaotic motion of classical dynamical systems [1]. Such irregular or chaotic dynamics is usually qualified as much more complex than the regular motion. But generically the majority of dynamical systems exhibit both regular and chaotic behavior at different energies and/or at different values of control parameters. The study of the transition from the regular to the chaotic limit should be done for parametric families of dynamic systems. Such a transition typically goes through several consecutive steps from completely regular motion to a partial breaking of integrals of motion [2]. At the same time, even at the level of completely regular motion for completely integrable systems, the complexity of a dynamical system can be increased through a sequence of bifurcations (in both classical and quantum cases) [3, 4]. Increasing complexity of a completely integrable classical (or quantum) dynamical systems can be most easily seen by analyzing the image of its energy–momentum map (or bifurcation diagram) [5, 6].

For a completely integrable classical Hamiltonian dynamical system, the energy–momentum map establishes the correspondence between common levels of energy and other integrals of motion and the values of these integrals that are in involution for a completely integrable classical system. We remind that two physical quantities are in involution in classical mechanics, if their Poisson bracket is zero. In quantum mechanics, the corresponding operators commute. The energy–momentum map has regular and critical points in the initial phase space of the classical Hamiltonian problem and regular and critical values in the space of values of integrals of motion. Inverse images of regular values are regular tori [7], while inverse images of critical values are various topologically different objects. Consequently, we can say that the energy–momentum map defines a fiber space with the base being the space of allowed values of integrals of motion and the fibers being the inverse images of the map. The set of critical values of the energy–momentum map defines in some sense the complexity of the integrable dynamical system. In order to characterize the complexity, we need to describe either the topology of singular (critical) fibers and their organization or to introduce some special characteristics of a family of regular fibers surrounding the above-mentioned singularities. This is exactly the moment when the Hamiltonian monodromy appears as a natural characteristic to describe the singularities of toric fibrations and the complexity of dynamical systems.

For quantum analogues, the system of common eigenstates of mutually commuting quantum operators that form the complete set of observables plays the role of toric fibration. Locally, the joint spectrum of mutually commuting operators forms a regular lattice in regions corresponding to regular toric fibrations of associated classical systems. At the same time, singularities of classical toric fibrations correspond to certain defects of the pattern representing the joint spectrum of quantum operators. We can equally try to describe the defect by analyzing the behavior of the locally regular lattice surrounding the defect. This leads us naturally to quantum monodromy that generalizes the classical Hamiltonian monodromy to quantum systems.

After this intuitive introduction giving initial ideas about relations between the complexity and monodromy of classical and quantum systems and singularities (or defects) of almost regular patterns associated with classical and quantum systems, we turn to a slightly more detailed description of the Hamiltonian monodromy along with more concrete examples of classical and quantum dynamical systems possessing monodromy. We will even try to generalize the characteristic discrete patterns with monodromy that arise naturally for quantum atomic and molecular systems and to look for generic universal patterns and their defects that appear in completely different scientific domains like botany, but that nevertheless can be considered as a result of the evolution of a complex dynamical system showing simple universal behavior.

The organization of this chapter is as follows. Section 8.2 is a basic introduction to classical Hamiltonian monodromy. Classical–quantum correspondence is used in Section 8.3 in order to explain associated quantum monodromy. Section 8.4 focuses on the interpretation of quantum monodromy in terms of defects of lattices of quantum states and deals with the associated elementary and complex defects. Possible generalizations of the monodromy concept, naturally arising during the analysis of concrete physical examples, are suggested in Section 8.5. Section 8.6 reviews recent progress in monodromy manifestations in time-dependent processes. Finally, Section 8.7 discusses perspectives of possible applications of the monodromy concept to very complex biological phenomena like plant morphogenesis, using as an example one particular but universal phenomenon: spiral phyllotaxis.

8.2
Hamiltonian Monodromy

The notion of "monodromy" is generally used in science, mainly in mathematics, in order to explain how some mathematical objects behave as they "go" around a singularity. This is a very general and imprecise description that can be made more concrete within the fiber space construction [5, 8].

Let $p : E \to B$ be a locally trivial fiber space with the base B. With each point $b \in B$, we associate a fiber $p^{-1}(b)$ and with each continuous path $\gamma : [0, 1] \to B$ in B with initial point $a = \gamma(0)$ and endpoint $b = \gamma(1)$, we can associate a homeomorphism of the fiber $p^{-1}(a)$ onto the fiber $p^{-1}(b)$. In a particular case of $a = b$, the path γ is a loop and we have a transformation of $F = p^{-1}(b)$ into itself. This transformation defined up to a homotopy is called the monodromy transformation. The transformation of fibers induces the transformation on the homology and cohomology spaces of F, which are also called a monodromy transformation.

In the context of classical Hamiltonian integrable dynamical systems, the monodromy transformation appears naturally as a transformation of a regular fiber associated with a closed path in the base space defined as a space of possible values of integrals of motion being mutually in involution [5, 9]. Image of the energy–momentum map (also known as bifurcation diagram) [6] enables us to clearly visualize for 2D integrable Hamiltonian systems the qualitative structure of the associated fiber space. Regular fibers in this case are regular tori [7] and the monodromy transformation can be described as a transformation of a fundamental group (or a homology group) of a fiber associated with closed paths in the base space. In a more classical mechanics way, we can speak about local action–angle variables defined on individual tori and their connection (evolution) with a family of tori associated with a loop in a base space. The fact that a loop goes around a singularity and is consequently noncontractible implies the absence of global action–angle variables and the presence of nontrivial monodromy transformation [9, 10].

The simplest situation in the case of completely integrable dynamical systems with two degrees of freedom corresponds to the presence of an isolated singular fiber (see Figure 8.1a) that can be surrounded by a loop in the base space going only through regular fibers [5, 11–14]. Such a situation is present, for example, in a number of very simple model mechanical problems like particle motion in an axially symmetric potential of "Mexican hat" or "champagne bottle" type: $V(r) = ar^4 - br^2$, with $a, b > 0$ [15, 16], or for spherical pendulum [5, 12, 17]. We can associate with each isolated singular fiber a closed path going around a corresponding isolated critical value in the base space. In its turn, this closed path is associated with the transformation of the first homology group of a regular fiber. This transformation can be naturally represented by a matrix with integer coefficients leading to integer monodromy. A typical isolated singular fiber that appears generically for Hamiltonian systems with two degrees of freedom is a pinched torus represented in Figure 8.2a. A pinched torus is obtained from a regular torus by shrinking one nontrivial circle to a point. Its geometrical view as an object in the three-dimensional ambient space can

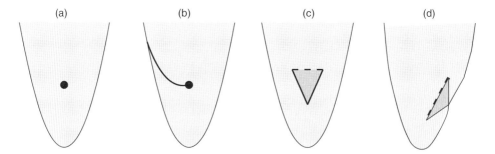

Figure 8.1 Typical images of the energy–momentum map for completely integrable Hamiltonian systems with two degrees of freedom in the case of (a) integer monodromy, (b) fractional monodromy, (c) nonlocal monodromy, and (d) bidromy. Values in the light-shaded area lift to single 2-tori; values in the dark-shaded area lift to two 2-tori. Dashed lines in figures (c) and (d) correspond to stratum formed by bitori.

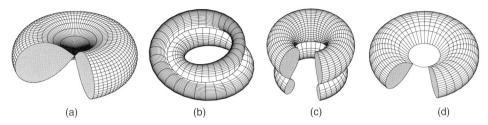

Figure 8.2 Two-dimensional singular fibers in the case of integrable Hamiltonian systems with two degrees of freedom. (a) Pinched torus. (b) Curled torus. (c) Bitorus. (d) Singular (cuspidal) torus.

give an impression that depending on the choice of vanishing cycle, the geometry of the pinched torus can be different. But this is just an artifact that is caused by plotting the pinched torus in 3D rather than in 4D. In fact, the choice of coordinates on the torus is ambiguous and this ambiguity is due to the $SL(2, Z)$ symmetry of a two-dimensional lattice. This leads, in particular, to the important fact that the matrix representation of monodromy is defined up to a $SL(n, Z)$ similarity transformation for n-dimensional problem.

It is possible that instead of one isolated singular fiber, the image of the energy–momentum map has a whole region with a complicated set of singular fibers surrounded nevertheless by a regular region. Figure 8.1c shows an example of an "island" that can be surrounded by a closed loop in the base space going through only regular fibers. The resulting monodromy transformation characterizes the whole region possessing singular fibers, and in such a case we can speak about nonlocal integer monodromy [18, 19]. One of the simplest reasons of the appearance of a nonlocal "island" like singularity is the formation of a second component in the image of the energy–momentum map due to the transformation of an isolated focus–focus singularity into an "island" through the Hamiltonian Hopf bifurcation [12] that is related to the presence of a family of singular fibers. Each generic member of such a family is named a bitorus. It is represented in Figure 8.2c. A singular (cuspidal) torus (see Figure 8.2d) is its limiting form corresponding to corners on the bifurcation diagram (Figure 8.1c) where the bitorus line ends.

A less trivial situation arises when the essentially singular fiber is not isolated but appears as a limiting case of a one-dimensional stratum of weakly singular fibers [12, 20, 21]. The presence of weakly singular fibers does not allow one to go around the essential singularity by transporting the basis cycles of the homology group of regular fibers. At the same time, it is now possible to study the continuous evolution of cycles for certain subgroups of the homology group. Such construction allows us to generalize the monodromy notion and introduce "fractional monodromy" [20–25]. An image of the energy–momentum map with a singularity leading to fractional monodromy is shown in Figure 8.1b. An example of weakly singular fibers, the so-called curled torus, is shown in Figure 8.2b.

Qualitative characterization of images of energy–momentum maps for classical integrable systems and especially of their generic possible evolution for families of integrable systems depending on one or several control parameters is important from the point of view of different generalizations. In spite of a very serious restriction due to integrability, the qualitative features, like monodromy, remain valid even for nonintegrable systems, which can be obtained by small nonintegrable perturbations [26]. The situation here is similar to the KAM theorem [7] ensuring that the majority of tori survive under a small perturbation leading from regular to chaotic classical motion.

8.3
Classical–Quantum Correspondence

Another important aspect of classical integrable systems with monodromy is the generalization from classical to quantum mechanics [13, 17, 27–29]. The quantum analogue of a classical integrable dynamical system is a quantum system possessing a mutually commuting set of operators corresponding to physical observables. Classical local action variables correspond in quantum mechanics to quantum numbers that label common eigenfunctions of mutually commuting operators. Representation of joint spectra of mutually commuting quantum operators naturally leads to locally regular lattices due to the simple quantization conditions imposed on local classical action variables.

Formal correspondence between classical integrable Hamiltonian systems and their quantum analogues helps to visualize the quantum monodromy [28]. Existence of local actions for an integrable system means that the joint eigenvalues of mutually commuting quantum operators corresponding to classical integrals of motion for a n-degree of freedom, completely integrable, dynamical system form a pattern that locally can be mapped onto a standard Z^n lattice [30, 31]. If the quantum commuting operators correspond directly to classical actions, the lattice formed by the joint spectrum of these operators represents a regular rectangular pattern because of the simple quantization rules for both action variables. A more typical situation corresponds to cases when only part of the classical integrals of motion are actions, while others, like energy, are not actions themselves, but can be locally transformed into actions by nonlinear (and in some sense small) transformations.

Figure 8.3 gives an example of classical–quantum correspondence for completely integrable problems with two degrees of freedom. The image of the classical energy–momentum map is shown in the space of values of integrals of motion: energy and first action. Figure 8.3a shows the regular part of the image of classical energy–momentum map together with points corresponding to joint eigenvalues of quantum mutually commuting operators. In order to see the natural correspondence of a discrete lattice with regular Z^2 lattice, an elementary cell of the lattice is chosen and displaced through the lattice along a closed path. It is evident that the initial and final cells coincide and all closed paths in this local regular region of the lattice are similar from this point of view. In more formal terms we can say that all closed paths in the regular simply connected region of the image of the energy–momentum map are homotopic to a point, that is, such closed paths could be shrunk to a point because all intermediate cells remain equivalent.

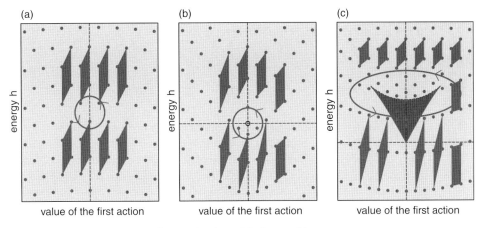

Figure 8.3 Quantum joint spectra for typical regions of the image of the energy–momentum map for some examples of integrable problems [32].

Figure 8.3b illustrates another possibility. The region of the classical energy–momentum map shown here has an isolated singularity, associated with a pinched torus. In any simply connected local region that does not include the singularity, the behavior of the quantum cell after going around a closed path is similar to that shown in Figure 8.3a. In contrast, if the closed path goes around the singularity, the final form of the elementary cell clearly differs from the initial form, thus indicating the presence of nontrivial monodromy [28]. The transformation between initial and final cells does not depend on the geometry of the closed path supposing that it goes only once around the singularity. The transformation between initial and final cells written in the matrix form gives the matrix representation of quantum monodromy. Naturally, the explicit form of the matrix depends on the choice of the lattice basis that is defined up to an $SL(2, Z)$ similarity transformation.

Figure 8.3c shows the image of the classical energy–momentum map with a region where two connected components exist in the inverse image. This curvilinear triangular region is represented in the figure by dark hashing. The correspondence between classical and quantum mechanics is more complicated for this example. Outside the region with two classical connected components, there is one lattice of quantum joint eigenvalues. Inside the "dark" region with two classical connected components of the inverse image, all quantum states can be approximately separated into two distinct leaves (neglecting the tunneling splitting that allows the coupling of quantum states associated with two disconnected classical components). Taking an elementary cell on the big leaf and going around the "island," we define a quantum nonlocal monodromy that characterizes the second leaf of the lattice and the way how two leaves join together. The vibrational structure of an LiCN molecule can be suggested as an example of a concrete molecular system that is reasonably well described by an integrable approximation that shows the presence of an island and a nonlocal nontrivial monodromy [19]. The second component exists as well for the quite similar example of an HCN molecule [33], but the closed path surrounding the second component cannot be constructed for the HCN model and consequently the HCN example is completely different from the LiCN one.

8.4
Lattices and Defects

The above-mentioned examples give an impression that patterns formed by the joint spectrum of mutually commuting quantum operators can be considered either as ideal periodic lattices (ideal crystals) or as periodic lattices with defects [30]. More generally, one can suppose that lattice defects of periodic crystals [34, 35] should correspond in some way to singular fibers of classical integrable dynamical systems. In fact, this analogy proves itself to be extremely useful, although even the simplest classical singularity associated with a pinched torus (the so-called focus–focus point) leads to a defect of the lattice of quantum eigenstates that has no straightforward analogy among typical crystal defects.

8.4.1
Elementary Monodromy Defects

In order to see the correspondence with solid-state defects, we analyze the pattern of quantum eigenstates for a two-degree-of-freedom integrable system possessing, in the classical limit, a focus–focus singularity. A typical classical image of the energy–momentum map together with common eigenstates of mutually commuting operators for corresponding quantum problem is shown in Figure 8.4. To demonstrate the presence of the defect, we make a cut starting at the singularity of the classical problem. There are a lot of different possibilities to realize such cut. We first make the cut along the $m = 0$, $E > 0$ ray. Such choice of the direction of a cut is named *eigenray* by M. Symington [14]. Its specificity can be easily explained by using

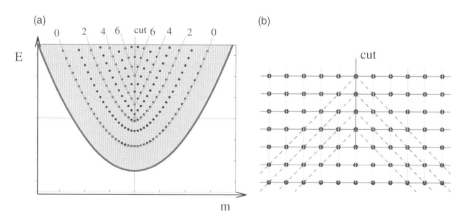

Figure 8.4 $1 : (-1)$ Resonant oscillator system with a cut along a *eigenray* $R = \{m = 0, E > 0\}$. (a) Representation in energy–angular momentum variables. (b) Representation in action variables.

Figure 8.4, where we have labeled eigenstates for each m by consecutive integers starting from zero. These integers play the role of quantum numbers associated with a second action that is defined locally in any simply connected region of regular values of the energy–momentum map. We see that this local action (quantum number) has the same value for each eigenstate located at the cut irrespective of whether we approach the cut from the right or from the left. At the same time, the first derivative $\partial E/\partial m$ (calculated for a set of states with the same quantum number corresponding to a second action) has a jump at the cut. Due to the presence of such a discontinuity of the first derivative, this cut was named "kink" line by M. Child [16]. It should be noted that the presence of a "kink" singularity is a purely artificial fact related to the multivaluedness of the action variables and to our choice of one leaf of the multi-valued function.

In fact, we can continue labeling eigenstates of the joint spectrum by continuing local action variables within $E > 0$ energy region from the $m > 0$ to the $m < 0$ domain. This will naturally give another labeling scheme that can be shown on the bifurcation diagram and that is associated with an alternative construction of a single valued function from an initially multivalued one. If we transform the (E, m) plot to new coordinates that we choose as two local actions (as it is shown in Figure 8.4b) in order to continue the line corresponding to one chosen value of local action across the $(m = 0, E > 0)$ ray, we need to change the direction after crossing the ray.

If we do cut along any other direction, the local action itself has a discontinuity at the cut. This situation is shown in Figure 8.5. Further transformation of the lattice with a cut to local action coordinates leads to a regular lattice with a certain part of this lattice removed and with the boundaries of this removed wedge being identified [21, 30, 31]. Figure 8.6 illustrates the construction of the defect associated with the simplest focus–focus singularity (single pinched torus) in the regular lattice by making a cut in the direction orthogonal to the "eigenray." Such construction is similar in spirit to the representation of dislocations and disclinations in solid-state physics by a cutting and gluing procedure. At the same time, the obtained defect differs from well-known constructions for dislocations and disclinations. We named

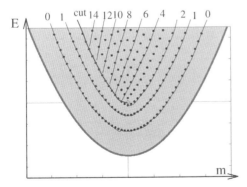

Figure 8.5 Cut leading to a discontinuity of the second local action.

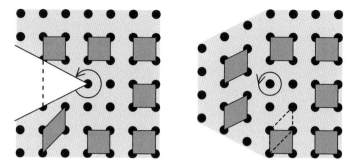

Figure 8.6 Construction of the 1 : (−1) lattice defect starting from the regular Z^2 lattice. Dark gray quadrangles show the evolution of an elementary lattice cell along a closed path around the defect point.

the defect shown in Figure 8.6 an "elementary monodromy defect." The most important feature of the suggested construction of an "elementary monodromy defect" is the linear dependence of the number of removed states on the value of the integral of motion. This property is related to the Duistermaat–Heckman theorem applied to the volume of the reduced phase space in classical mechanics [27, 36, 37].

8.4.2
Fractional Monodromy Defects

The geometrical construction proposed for the description of a monodromy defect immediately allows several alternative ways of generalization. The first possibility is the construction of less trivial elementary defects, like recently introduced fractional monodromy defects [20, 21]. We illustrate construction of such a defect in Figure 8.7. The idea of relevance of such a specific defect to patterns formed by joint eigenvalues of mutually commuting operators came from the typical dependencies of the number of states in multiplets or polyads on the quantum number characterizing these

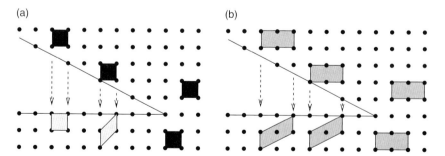

Figure 8.7 Construction of 1 : 2 rational lattice defect. (a) Elementary cell does not pass unambiguously. (b) Double cell passes unambiguously.

polyads or multiplets. This question is again related to the Duistermaat–Heckman approach describing the evolution of the reduced phase spaces with the corresponding integral of motion value.

The important difference between the removed wedge associated with elementary "integer" monodromy (represented in Figure 8.6) and the removed wedge in the case of a fractional monodromy defect (see Figure 8.7) is in the number of the removed states considered as a function of the integral of motion. In the integer case, this function is linear (or polynomial in the higher dimensional case) [37], while in the fractional case the function is a "quasi-polynomial" [38], that is, it includes an oscillatory part. In a particular example shown in Figure 8.7, we have just modulo 2 oscillations. In an equivalent way, it is possible to say that we need to consider separately the sublattices with even and odd m values, each possessing an integer "elementary monodromy defect." The geometrical consequence is the impossibility for an elementary cell to go unambiguously through the cut. Depending on the position after crossing the cut, the elementary cell takes one of two different geometrical forms shown in Figure 8.7a. In contrast, if we use a double cell, the result of the cell transformation after making a closed trip around the essential singularity and crossing the cut only once is independent of the place where the cell crosses the cut. The price for that are the fractional coefficients that appear in the monodromy matrix written for the elementary cell. We can formally use the monodromy matrix for an elementary cell in the regular region even if this elementary cell itself cannot cross the singular cut.

An example of a simple model problem showing the presence of a fractional monodromy and the associated pattern of common eigenstates of two mutually commuting operators is shown in Figure 8.8 [39].

The corresponding dynamical system is constructed by taking two angular momentum operators $\boldsymbol{N} = (N_x, N_y, N_z)$, $\boldsymbol{S} = (S_x, S_y, S_z)$ interacting in a nontrivial nonlinear way between themselves and with an external field. The model

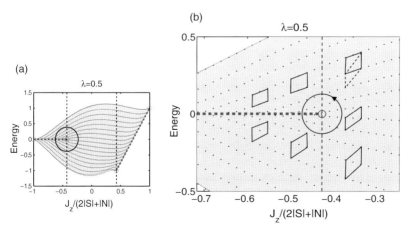

Figure 8.8 Fractional monodromy for two angular momenta coupling model. Image of the energy–momentum map is given for Hamiltonian (8.1) with $\lambda = 0.5$.

Hamiltonian is given as

$$H_\lambda = \frac{1-\lambda}{|S|} S_z + \lambda \left(\frac{1}{|S||N|} S_z N_z + \frac{1}{2|S||N|^2} \left(N_-^2 S_+ + N_+^2 S_- \right) \right), \tag{8.1}$$

where standard notation is used for ladder operators $N_\pm = N_x \pm i N_y$, $S_\pm = S_x + i S_y$ and $|N| = \sqrt{N_x^2 + N_y^2 + N_z^2}$, $|S| = \sqrt{S_x^2 + S_y^2 + S_z^2}$. This model generalizes the most trivial model of angular momenta coupling leading to the appearance of a focus–focus singularity in classical mechanics and an integer quantum monodromy in the corresponding quantum problem [28, 40–42]. It is interesting to note that the simplest quantum problem, namely, the hydrogen atom, in the presence of external electric and magnetic fields leads to integrable approximations for certain values of field parameters, which show the presence of fractional monodromy [43–45].

8.4.3
Monodromy–Defect Correspondence

Another way of generalizing the possible patterns typical for lattices formed by joint spectrum of mutually commuting operators is to describe more complicated defects that can be present for 2D lattices from one side and for higher dimensional lattices from another side. We restrict ourselves here with 2D case only.

First of all it should be reminded that if we want to characterize the defect of a quantum lattice (or the singularity of a classical toric fibration) by its monodromy, we must introduce in some sense a certain equivalence relation between different defects (singularities). The most detailed description of a singularity of classical toric fibration and of the associated quantum state lattice can be done by indicating the explicit transformation of the basis of the first homology group of regular fibers. In such a case, two defects characterized by the presence of one vanishing cycle become different if the vanishing cycles themselves are different. Naturally, the monodromy matrices associated with these two defects and written on the same basis of cycles (or on the same lattice basis) are different. At the same time, these matrices belong to the same class of conjugated matrices within the $SL(2, Z)$ transformation group responsible for the basis transformation of the 2D lattice. Thus, the monodromy matrices should be considered equivalent if they belong to the same class of conjugated elements of the $SL(2, Z)$ group. To characterize the class of conjugated elements, we first use the trace of the matrix. The matrices $M \in SL(2, Z)$ are named parabolic, elliptic, or hyperbolic, if their traces equal ± 2 (for parabolic), $\pm 1, 0$ (for elliptic), or $> |2|$ (for hyperbolic). A more detailed classification of parabolic matrices into classes of conjugated elements needs additional information, like a nondiagonal element of some standard "normal form" of the parabolic matrices. Especially interesting is the importance of the sign of the nondiagonal element indicating that, for example, two quite simple monodromy matrices

$$\begin{pmatrix} 1 & 0 \\ 1 & 1 \end{pmatrix}, \quad \begin{pmatrix} 1 & 0 \\ -1 & 1 \end{pmatrix}$$

belong to different classes of conjugated elements within the $SL(2, Z)$ group. We name these matrices positive and negative elementary monodromy matrices. It is important to note that only one class corresponds to an elementary toric singularity, the pinched torus. Its representation as a defect of a regular lattice corresponds to removing a wedge from the lattice and to identifying the wedge boundaries. It is also important to note that using several elementary monodromy matrices that belong to the same class of conjugated elements of $SL(2, Z)$ but can be reduced to the normal form in a different lattice basis, it is possible to construct an arbitrary $SL(2, Z)$ matrix representing a cumulative monodromy effect [46, 47].

At the same time, the monodromy (i.e., the class of conjugated elements of $SL(2, Z)$ group) does not define in a unique way the classical singularity and the pattern formed by the joint spectrum of commuting quantum operators [46, 47]. For example, 12 specially constructed elementary singularities (each described by a matrix conjugated to the simplest monodromy matrix and represented by defects associated with removing a wedge from the lattice) can lead to global trivial monodromy. This is a simple consequence of the $SL(2, Z)$ group structure [48, 49]. Nontriviality of a global singularity nevertheless is clearly seen through the transformation of the elementary cell going around all singularities on the lattice of mutual quantum eigenstates. The cell realizes the 2π rotation around itself while going along a closed path surrounding these 12 elementary singularities. It is important that the sign of the rotation of the elementary cell is well defined and corresponds to removing wedges from the lattice [31].

As another example, it is possible to realize the negative "elementary monodromy" by using 11 elementary positive monodromy defects. In such a case, the rotation of an elementary cell while going around this defect can again be characterized as a "positive" and being almost overall 2π rotation in spite of the fact that a much more simpler construction of the elementary "negative defect" can be formally done by inserting into the lattice a wedge corresponding to one elementary monodromy defect instead of removing from the lattice 11 wedges corresponding to elementary defects as in the case of "positive defects."

The notion "elementary monodromy" that we use is from one side due to the simplicity of the matrix representation but from another side due to the simplicity of the topology of the singular classical fiber responsible for such monodromy. The geometric monodromy theorem states that the presence of an isolated singly pinched torus leads to the elementary positive monodromy of an associated toric fibration [5, 11]. A singly pinched torus can be alternatively described as a sphere with one transversal positive self-intersection point. In order to understand this statement better, it should be reminded that in four-dimensional space, the two-dimensional surfaces generically intersect via isolated points and the simplest model of a pinched point of a torus corresponds to the intersection of two 2D planes ($x = 0$, $y = 0$) and ($z = 0$, $w = 0$) in 4D $\{x, y, z, w\}$ space. The positivity of the self-intersection point means that a 4D frame constructed from two 2D frames transported from a regular point on the surface to the singular point of self-intersection via two nonequivalent paths gives positive volume.

The representation of a negative "elementary monodromy" in terms of 11 elementary positive singularities shows in some sense the complicated nature of the "elementary negative defect." At the same time, simple topological arguments suggest, instead of a singly pinched torus as a possible candidate for an alternative simple defect, the singular fiber that is a sphere with one transversal negative self-intersection point [50]. The objection to appearance in Hamiltonian systems of such a topologically allowed defect comes from the deformation arguments. Supposing that the "negative" defect is an elementary one, the small deformation of this singular fiber removing the singular point should lead to a fiber from the neighborhood filled by regular fibers. At the same time, a small deformation of a singular fiber that is a sphere with one transversal negative self-intersection point should lead to a regular fiber that is a nonorientable surface (Klein bottle) instead of a regular torus (Figure 8.9, 8.10).

In order to see why this nonorientable surface cannot be another nonorientable surface, for example, real projective plane RP_2, rather than the Klein bottle (Figure 8.9), we can study what happens with a pinched RP_2 if we cut it through the pinch point. One can verify that a singly pinched RP_2, after cutting through the pinch point, becomes again RP_2 rather than sphere.

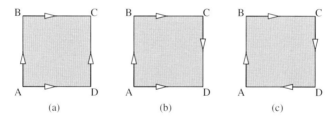

Figure 8.9 Construction of two-dimensional surfaces by the identification of the opposite ends of a square, preserving the directions indicated by arrows. (a) Identification of AB with DC and AD with BC gives torus. (b) Identification of AB with CD and AD with BC gives Klein bottle. (c) Identification of AB with CD and BC with DA gives real projective plane RP_2.

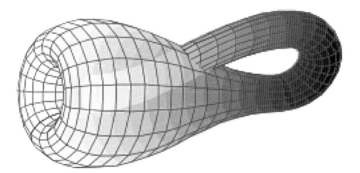

Figure 8.10 Representation of a Klein bottle in a 3D space as a surface with self-intersection.

In contrast, if we prepare a singly pinched Klein bottle and cut it through the pinch point, the result may be different. It is important to note that the Klein bottle can be pinched in two nonequivalent ways: either by shrinking to zero the cycle that is a generator of the infinite group, or by shrinking to zero a generator of a Z_2 group. Cutting a pinched Klein bottle through the pinch point that corresponds to a vanishing cycle being a Z_2 generator leads to a 2D sphere S_2. At the same time, cutting pinched Klein bottle through the pinch point that corresponds to a vanishing cycle being a Z generator of a group of integers results in an RP_2 surface. Thus, appearance of a sphere with one negative self-intersection point as a generic singular fiber assumes that regular fibers should be Klein bottles rather than regular tori and the singularity should be associated with the vanishing of the Z_2 generator. It is probably useful to note here that the Klein bottle itself can be considered as a critical fiber of toric fibration associated with the double covering of a Klein bottle by a torus. Although a Klein bottle is known to be a generic critical fiber for toric foliations, this critical fiber rarely appears in applications [6]. What kind of integrable systems can lead to generic fibers of Klein bottle type remains an open problem.

8.5
Multicomponent Energy–Momentum Map, Bidromy, and Others

Another generalization of monodromy is based on the analysis of images of energy–momentum maps with several components. A rather complicated example of multicomponent maps arises even for such a naturally simple integrable model as a Manakov top [51]. A much more simpler example of the appearance of a second component was shown in Figure 8.1c, where the second component appears as a result of a Hamiltonian Hopf bifurcation [12, 52] leading to transformation of an isolated focus–focus singular point into a second leaf attached to the main leaf through a family of singular fibers, named bitori (see Figure 8.2c). Such creation of the second leaf evidently cannot modify the nontrivial monodromy associated with an initial isolated singular point and consequently the nonlocal monodromy is associated with the closed path surrounding the whole second leaf on the image of the energy–momentum map. More precisely, we should say that the closed path goes around bitorus stratum responsible for joining two components into one. Examples of such a transformation are well known in different molecular examples like a hydrogen atom in fields [52, 67] or an LiCN [19, 33] molecule. It should also be noted that the appearance of the second leaf on the image of the energy–momentum map could also result from the fold-type catastrophe [32]. In such a case, the second component typically appears within the regular region of the image of the energy–momentum map and consequently the monodromy transformation associated with the closed path surrounding the second leaf should remain trivial in this case.

Another possibility of getting the second leaf appears in a rather different situation associated in fact with the self-overlapping of the same leaf (see Figure 8.1d). The organization of the image of the energy–momentum map can be explained by using a schematic "unfolding" procedure represented in Figure 8.11, which can be more

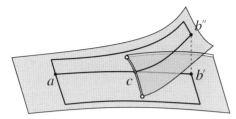

Figure 8.11 Self-overlapping of a lower cell of the unfolding image of the energy–momentum map.

accurately explained by introducing notions of upper and lower cells for singular toric fibrations as it is done in detail in Ref. [21].

The specificity of the situation is due to the fact that a certain region of the image of the energy–momentum map has two regular tori as an inverse image (points b' and b'' in Figure 8.11). At the same time, we can choose a continuous family of *regular* tori allowing deformation of one torus in this region into another. The possibility of such a deformation leads to an unambiguous definition of three tori lying close to a bitorus fiber (point c in Figure 8.11) on three locally different leaves of the image of the energy–momentum map. This construction gives the possibility to define the crossing of a bitorus line in such case. The tentative definition of corresponding splitting of cell and path when crossing the bitorus line and fusion of bipath into one path along with two cell fusion was named *bidromy* [53]. The bidromy phenomenon was initially illustrated on a rather complicated example of a three-degree-of-freedom system with special resonance 1 : 1 : 2 [53, 54], but essentially the same qualitative behavior can be observed for a two-degree-of-freedom dynamical system. An example of such behavior was found even in simple quantum systems, for example, a hydrogen atom in the presence of electric and magnetic external fields [43, 45].

8.6
Time Evolution and Monodromy

Up until now, only a "static", in some sense, manifestation of Hamiltonian monodromy, namely, special arrangements of joint spectrum of mutually commuting operators, has been discussed. A quite interesting direction of further investigation is related with the analysis of monodromy manifestations during time-dependent processes. The general imprecise idea is to realize the time-dependent evolution of the dynamical system corresponding in some sense to going along a closed path surrounding a singularity in the energy–momentum space. The initial question in such a construction is about what should be observed and what fingerprints of monodromy could be found. The first step in this direction was made by Delos *et al.* [55, 56] using very simple toy problem, namely, the motion of a single particle in an axially symmetric billiard with a parabolic barrier potential. This problem possesses monodromy in its stationary Hamiltonian formulation. In order to see the nontrivial dynamic effect of monodromy, one needs to follow the evolution of a family of particles and to choose a special time-dependent perturbation that allows the change of values of the integral of motion in, say, an adiabatic way. The dynamic manifestation of monodromy consists in the nontrivial topological modification of an initial spatial distribution of particles after following a closed path in the energy--momentum space and returning to the initial values of integrals of motion. The most difficult step in the realization of such a time-dependent processes is to find the precise form of the time-dependent perturbation that satisfies all theoretically imposed assumptions on the form of perturbation and to realize it practically.

8.7
Perspectives

The notion of monodromy can be related to problems that are quite far from the classical Hamiltonian integrable systems or model quantum molecules. The idea of such a generalization is based on the relation between defects of regular patterns and the monodromy. Namely, many defects of regular lattices that appear in solid-state physics can be considered as a cumulative result of a number of elementary monodromy defects and can be treated as some complicated nonelementary defects from the point of view of monodromy defects. It is clear that the choice of "elementary bricks" is not unique. Even though the mathematical description of defects in solid-state physics and in toric fibrations related to dynamical Hamiltonian systems, or in other models turns out to be similar, the relevance of these mathematical constructions should be confronted with physical reality.

From the physical point of view, the origin of defects in solids is due to imperfection of the crystal growth. Some of the 2D point defects like disclinations have a natural description in terms of several elementary monodromy defects. Some others, like vacations, have nothing to do with monodromy, because they are not related to the topology of the lattice. We can try to generalize the mathematical description of defects in terms of elementary monodromy defects and to look from this point of view for typical singularities (defects) of almost regular patterns appearing in different domains. The general idea behind this is to find interpretation of defects in terms of natural "elementary ones" and to specify the generic most frequently appearing defects and to find a possible explanation of their appearance.

Regular patterns with defects can appear not only in solid state, with each point being associated with an atom or molecule, but also in more complex systems like plants with regular patterns being associated with leaves or seeds, reflecting the morphogenesis or the plant development. The most striking example of such a regular pattern formation is the phyllotaxis, intriguing scientists – Leonardo da Vinci, Kepler, Bravais, Turing, Coxeter, to name a few [57] – working in different fields even quite far from biology.

The phenomenon of phyllotaxis describes the morphology of many botanical objects. It exists in the arrangement of repeated units such as leaves around a stem in various plants, seeds of a pine cone or of a sunflower, scales of a pineapple, and so on. The most widely known is the spiral phyllotaxis associated in a major part of cases with lattices formed by left- and right-hand spirals whose numbers are found to be consecutive in the Fibonacci sequence $1, 1, 2, 3, 5, 8, \ldots, a_k, a_{k+1}, a_{k+2} = a_k + a_{k+1}, \cdots$. The interdisciplinary character of the phyllotaxis phenomenon is clearly seen in the example of pattern formation by drops of ferro fluids in a magnetic field [58] or by flux lattices in superconductors [59].

The enormous literature devoted to the study and to the explanation of the universal behavior of botanical patterns mainly deals with a peculiar presence of Fibonacci numbers and chemical regulation of their presence (see Refs [57, 60–64] and references therein). Characterization of the resulting pattern from the point of view of the singularity responsible for the pattern formation and its monodromy has, to our knowledge, not been described earlier in the literature. The appearance of defects within the spiral pattern and modifications of (n_1, n_2) into (n'_1, n'_2) patterns has been discussed on several occasions [65], but the most essential persistent feature of global organization of spirals due to the characteristic relation between a locally regular lattice and its nontrivial behavior around the growing center has not been related to a monodromy-like notion.

I would like to demonstrate here the manifestation of a phyllotaxis monodromy on the sunflower example and to formulate several questions about universality of the patterns, associated defects, and relevance of such defects to evolution processes not only in botanic but also in other fields of science.

First of all, taking an example of sunflower with 34 left and 55 right eye-guided spirals – parastichies in biological language – (see Figure 8.12), we note that locally regular lattice can be constructed by explicitly plotting spirals (see Figure 8.13). In any local simply connected region, this lattice can be easily transformed into a regular lattice, but globally it has an easily seen defect – the apex. In order to see the nontrivial effect of the apex region on the lattice, we take the elementary cell in any regular part of the lattice and move it step by step along a closed path surrounding the apex region

Figure 8.12 Sunflower with 55 right spirals (parastichies) indicated by additional lines to guide the eyes.

Figure 8.13 The same sunflower as in Figure 8.12, but now 34 left spirals are indicated.

(see Figure 8.14). At each step the local regular structure allows us to move the elementary cell unambiguously, even though the lattice itself and the corresponding elementary cell could be chosen in a different way because the basis of the lattice is defined as usual up to an $SL(2, Z)$ similarity transformation.[1] It is important to keep the vertices of the cell labeled at each step of the lattice displacement. We use letters a and b in Figure 8.14.

It is easy to verify that after following a closed path that does not go around the apex, the elementary cell returns to its initial position. Apparently the same situation occurs after going along a closed path surrounding the apex. But the principal difference is that in this case (shown in Figure 8.14) the cell returns to its original position after making a 2π self-rotation around axis passing through the center of the cell in an orthogonal to the cell plane direction. From the point of view of Hamiltonian monodromy, the comparison of the initial cell and the final cell can be expressed by an identity matrix that is associated with a trivial monodromy. At the same time, the closed path is evidently noncontractible and in order to characterize the singularity (or the defect) responsible for this nontriviality, we need to add another topological invariant associated with the closed path, namely, the self-rotation number that can be positive, negative, or zero. According to the earlier formulated correspondence between a cell transportation around a defect and the "monodromy defect" construction (by removing or inserting a wedge), the positive numbers of self-rotation correspond to defects with removing wedge. These defects correspond to typical singularities (focus–focus) for Hamiltonian systems. The direction of self-rotation of the cell is well defined for an arbitrary number of focus–focus singularities in Hamiltonian systems [30, 46, 47]. This direction is related to the fact that the corresponding defect is produced by removing a certain number of wedges from the

---

1) The corresponding ambiguity in the choice of parastichies was equally mentioned in literature discussing the geometrical aspects of phyllotaxis [66].

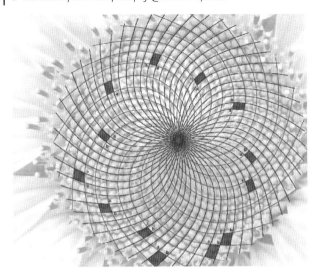

Figure 8.14 "Sunflower lattice" formed by left and right spirals shown in Figures 8.12 and 8.13. The transformation of an elementary cell of this lattice along the closed path surrounding the central singularity shows the presence of monodromy, related to a 2π self-rotation of the elementary cell (see text).

lattice. It is important to note that the direction of "positive" self-rotation should be defined only after specifying the direction of the closed path going around the singularity. The rotation can be defined positive if the cell turns from the $\mathbf{v}_1$ to $\mathbf{v}_2$ in the shortest way, where $\mathbf{v}_1$ is a vector defining the direction of the path surrounding the singularity, and $\mathbf{v}_2$ is a vector joining the cell and the singularity. More formally, the triple product $(\mathbf{v}_1, \mathbf{v}_2, \mathbf{R}) > 0$ should be positive, where $\mathbf{R}$ is the axial vector giving the self-rotation of the cell.

In a more intuitive way we can say that the positive choice of the cell's self-rotation coincides with the sense of rotation of a wheel (representing the cell) turning around another wheel (representing the singularity) if the "cell" rolls around the "singularity" without slipping. Unfortunately, in order to get in such a representation the value of the self-rotation angle to be equal to 2π, the radius of a wheel representing the singularity should be chosen to be equal to 0.

The choice of the sign of the monodromy defect observed for a wide range of botanic patterns seems to be rather fundamental property similar in the spirit to the left–right asymmetry and time irreversibility. A number of interesting questions naturally arise provoked by this supposition. Can the evolution of the plants be modeled by a dynamical system with the source being associated with a generic singularity characterized by an identity monodromy and a positive 2π self-rotation? Does the sign of self-rotation reflect specific properties of the system? We end with an even more general tentative speculation: Is the generic singularity associated with irreversible time evolution (growing process) always characterized by a trivial (identity) monodromy with positive 2π self-rotation? Is it possible to apply this conjecture to the evolution of our Universe?

References

1. Percival, I.C. (1973) Regular and irregular spectra. *J. Phys. B*, **6**, L229–L232.
2. Gutzwiller, M. (1990) *Chaos in Classical and Quantum Mechanics*, Springer.
3. Sadovskií, D.A. and Zhilinskií, B.I. (1993) Group-theoretical and topological analysis of localized rotation–vibration states. *Phys. Rev. A*, **47**, 2653–2671.
4. Zhilinskii, B.I. (2009) Quantum bifurcations, in *Encyclopedia of Complexity and System Science* (ed. R. Meyers), Springer, New York.
5. Cushman, R.H. and Bates, L. (1997) *Global Aspects of Classical Integrable Systems*, Birkhäuser, Basel.
6. Bolsinov, A.V. and Fomenko, A.T. (2004) *Integrable Hamiltonian Systems: Geometry, Topology, Classification*, Chapman & Hall/CRC, London.
7. Arnold, V.I. (1989) *Mathematical Methods of Classical Mechanics*, Springer, New York.
8. Nakahara, M. (1990) *Geometry, Topology and Physics*, IOP Publishing, Bristol.
9. Duistermaat, J.J. (1980) On global action angle coordinates. *Commun. Pure Appl. Math.*, **33**, 687–706.
10. Nekhoroshev, N.N. (1972) Action–angle variables and their generalizations. *Trans. Moscow Math. Soc.*, **26**, 180–198.
11. Matveev, V.S. (1996) Integrable Hamiltonian systems with two degrees of freedom. The topological structure of saturated neighborhoods of points of focus–focus and saddle–saddle type. *Sb. Math.*, **187**, 495–524.
12. Efstathiou, K. (2004) *Metamorphoses of Hamiltonian Systems with Symmetry*, Lecture Notes in Mathematics, vol. 1864, Springer, Heidelberg.
13. Vũ Ngọc, S. (1999) Quantum monodromy in integrable systems. *Commun. Math. Phys.*, **203**, 465–479.
14. Symington, M. (2001) Four dimensions from two in symplectic topology, in *Topology and Geometry of Manifolds*, Athens, GA, Proc. Symp. Pure Math., 71, 153–208, 2003.
15. Bates, L. (1991) Monodromy in the champagne bottle. *J. Appl. Math. Phys.*, **42**, 837–847.
16. Child, M. (2007) Quantum monodromy and molecular spectroscopy. *Adv. Chem. Phys.*, **136**, 39–95.
17. Cushman, R.H. and Duistermaat, J.J. (1988) The quantum mechanical spherical pendulum. *Bull. Am. Math. Soc.*, **19**, 475–479.
18. Sadovskií, D.A. and Zhilinskií, B.I. (2006) Quantum monodromy, its generalizations and molecular manifestations. *Mol. Phys.*, **104**, 2595–2615.
19. Joyeux, M., Sadovskii, D.A., and Tennyson, J. (2003) Monodromy of LiNC/NCLi molecule. *Chem. Phys. Lett.*, **382**, 439–442.
20. Nekhoroshev, N.N., Sadovskií, D.A., and Zhilinskií, B.I. (2002) Fractional monodromy of resonant classical and quantum oscillators. *C.R. Acad. Sci. I*, **335**, 985–988.
21. Nekhoroshev, N.N., Sadovskií, D.A., and Zhilinskií, B.I. (2006) Fractional Hamiltonian monodromy. *Ann. Henri Poincaré*, **7**, 1099–1211.
22. Efstathiou, K., Cushman, R.H., and Sadovskii, D.A. (2007) Fractional monodromy in the 1 : −2 resonance. *Adv. Math.*, **209**, 241–273.
23. Giacobbe, A. (2008) Fractional monodromy: parallel transport of homology cycles. *Diff. Geom. Appl.*, **26**, 140–150.
24. Nekhoroshev, N.N. (2007) Fractional monodromy in the case of arbitrary resonances. *Sb. Math.*, **198**, 383–424.
25. Sugny, D., Mardešić, P., Pelletier, M., Jebrane, A., and Jauslin, H.R. (2008) Fractional Hamiltonian monodromy from a Gauss–Manin monodromy. *J. Math. Phys.*, **49**, 042701.
26. Broer, H.W., Cushman, R.H., Fassò, F., and Takens, F. (2007) Geometry of KAM-tori for nearly integrable Hamiltonian systems. *Ergod. Th. Dynam. Sys.*, **27**, 725–741.
27. Vũ Ngọc, S. (2007) Moment polytopes for symplectic manifolds with monodromy. *Adv. Math.*, **208**, 909–934.

28 Sadovskií, D.A. and Zhilinskií, B.I. (1999) Monodromy, diabolic points, and angular momentum coupling. *Phys. Lett. A*, **256**, 235–244.

29 Grondin, L., Sadovskii, D.A., and Zhilinskii, B.I. (2002) Monodromy in systems with coupled angular momenta and rearrangement of bands in quantum spectra. *Phys. Rev. A*, **142**, 012105.

30 Zhilinskii, B.I. (2006) Hamiltonian monodromy as lattice defect, in *Topology in Condensed Matter, Springer Series in Solid State Sciences*, vol. 150 (ed. M. Monastyrsky), Springer, pp. 165–186.

31 Zhilinskii, B.I. (2005) Interpretation of quantum Hamiltonian monodromy in terms of lattice defects. *Acta Appl. Math.*, **87**, 281–307.

32 Dhont, G. and Zhilinskii, B.I. (2008) Classical and quantum fold catastrophe in the presence of axial symmetry. *Phys. Rev. A*, **78**, 052117.

33 Efstathiou, K., Joyeux, M., and Sadovskii, D.A. (2003) Global bending quantum number and the absence of monodromy in the HCN↔CNH molecule. *Phys. Rev. A*, **69**, 032504.

34 Mermin, N.D. (1979) The topological theory of defects in ordered media. *Rev. Mod. Phys.*, **51**, 591–648.

35 Kleman, M. (1983) *Points, Lines, and Walls*, John Wiley & Sons, Inc., New York.

36 Duistermaat, J.J. and Heckman, G.J. (1982) On the variation in the cohomology of the symplectic form of the reduced phase space. *Inv. Math.*, **69**, 259–268.

37 Guillemin, V. (1994) *Moment Map and Combinatorial Invariants of Hamiltonian Tn-Spaces, Progress in Mathematics*, vol. 122, Birkhäuser, Boston.

38 Stanley, R.P. (1986) *Enumerative Combinatorics*, vol. 1, Wadsworth & Brooks/Cole, Montrey, CA, Chapter 4.4.

39 Hansen, M.S., Faure, F., and Zhilinskii, B.I. (2007) Fractional monodromy in systems with coupled angular momenta. *J. Phys. A*, **40**, 13075.

40 Pavlov-Verevkin, V.B., Sadovskii, D.A., and Zhilinskii, B.I. (1988) On the dynamical meaning of the diabolic points. *Europhys. Lett.*, **6**, 573–578.

41 Faure, F. and Zhilinskii, B.I. (2000) Topological Chern indices in molecular spectra. *Phys. Rev. Lett.*, **85**, 960–963.

42 Faure, F. and Zhilinskii, B.I. (2002) Topologically coupled energy bands in molecules. *Phys. Lett. A*, **302**, 242–252.

43 Efstathiou, K., Sadovskii, D.A., and Zhilinskii, B.I. (2007) Classification of perturbations of the hydrogen atom by small static electric and magnetic fields. *Proc. R. Soc. Lond. A*, **463**, 1771–1790.

44 Efstathiou, K., Lukina, O.V., and Sadovskii, D.A. (2008) Most typical 1 : 2 resonant perturbation of the hydrogen atom by weak electric and magnetic fields. *Phys. Rev. Lett.*, **101**, 253003.

45 Efstathiou, K., Lukina, O.V., and Sadovskii, D.A. (2009) Complete classification of qualitatively different perturbations of the hydrogen atom in weak near-orthogonal electric and magnetic fields. *J. Phys. A*, **42**, 055209.

46 Cushman, R.H. and Vũ Ngọc, S. (2002) Sign of the monodromy for Liouville integrable systems. *Ann. H. Poincaré*, **3**, 883–894.

47 Cushman, R.H. and Zhilinskii, B.I. (2002) Monodromy of two degrees of freedom Liouville integrable system with many focus–focus singular points. *J. Phys. A*, **35**, L415–L419.

48 Serre, J.-P. (1973) *A Course in Arithmetics*, Springer, New York.

49 Rankin, R.A. (1977) *Modular Forms and Functions*, Cambridge University Press, Cambridge.

50 Matsumoto, Y. (1985) Torus fibrations over the two sphere with the simplest singular fibers. *J. Math. Soc. Japan*, **37**, 605–636.

51 Sinitsyn, E. and Zhilinskii, B. (2007) Qualitative analysis of the classical and quantum Manakov top. *SIGMA*, **3**, 046.

52 Efstathiou, K., Cushman, R.H., and Sadovskii, D.A., (2004) Hamiltonian Hopf bifurcation of the hydrogen atom in crossed fields. *Physica D*, **194**, 150–174.

53 Sadovskii, D. and Zhilinskii, B. (2007) Hamiltonian systems with detuned 1 : 1 : 2 resonance. Manifestations of bidromy. *Ann. Phys.*, **322**, 164–200.

54 Giacobbe, A., Cushman, R.H., Sadovskii, D.A., and Zhilinskii, B.I. (2004)

Monodromy of the quantum 1 : 1 : 2 resonant swing–spring. *J. Math. Phys.*, **45**, 5076–5100.

55 Delos, J.B., Dhont, G., Sadovskii, D.A., and Zhilinskii, B.I. (2008) Dynamical manifestation of Hamiltonian monodromy. *Europhys. Lett.*, **83**, 24003.

56 Delos, J.B., Dhont, G., Sadovskii, D.A., and Zhilinskii, B.I. (2009) Dynamical manifestations of Hamiltonian monodromy. *Ann. Phys.*, **324**, 1953–1982.

57 Adler, I., Barabe, D., and Jean, R.V. (1997) A history of the study of phyllotaxis. *Ann. Bot.*, **80**, 231–244.

58 Douady, S. and Couder, Y. (1992) Phyllotaxis as a physical self-organized growth process. *Phys. Rev. Lett.*, **68**, 2098–2101.

59 Levitov, L.S. (1991) Phyllotaxis of flux lattices in layered superconductors. *Phys. Rev. Lett.*, **66**, 224–227.

60 Jean, R.V. (1994) *Phyllotaxis: A Systematic Study in Plant Morphogenesis*, Cambridge University Press, Cambridge.

61 Turing, A. (1952) The chemical basis of morphogenesis. *Phil. Trans. R. Soc. B*, **237**, 37–72.

62 Reinhardt, D., Stieger, E., Mandel, T., Baltensperger, K., Bennett, M., Traas, J., Frimi, J., and Kuhlemeler, C. (2003) Regulation of phyllotaxis by polar auxin transport. *Nature*, **426**, 255–260.

63 Jönsson, H., Heisler, M.G., Shapiro, B.E., Meyerowitz, E.M., and Mjolsness, E. (2006) An auxin-driven polarized transport model for phyllotaxis. *Proc. Natl. Acad. Sci. USA*, **103**, 1633–1638.

64 Smith, R.S., Guyomarc'h, S., Mandel, T., Reinhardt, D., Kuhlemeier, C., and Prusinkiewicz, P. (2006) A plausible model of phyllotaxis. *Proc. Natl. Acad. Sci. USA*, **103**, 1301–1306.

65 Shipman, P.D. and Newell, A.C. (2005) Polygonal planforms and phyllotaxis on plants. *J. Theor. Biol.*, **236**, 154–197.

66 Coxeter, H.C.M. (1961) *Introduction to Geometry*, John Wiley & Sons, Inc., New York.

67 Cushman, R.H. and Sadovskii, D.A. (2000) Monodromy in the hydrogen atom in crossed fields. *Physica D*, **142**, 166–196.

9
Dynamics in Materials Science

Gérard A. Maugin and Martine Rousseau

9.1
Introduction

Dynamics or the science of motion under the action of forces is viewed as a part of mechanics, the other part being statics or the science of equilibrium. In a very sketchy view, mechanics itself has developed in several identifiable stages: first, the mechanics of points and rigid bodies (seventeenth and eighteenth centuries), and then the mechanics of fluids (eighteenth century) and deformable solids (essentially nineteenth century), the mechanics of continua being understood as a synthesis of the two, under the unifying umbrella of thermodynamics (twentieth century). It is only in the last few decades of the twentieth century that a new avatar of continuum mechanics in the form of the *mechanics of materials* began to appear. This developed from contacts with solid-state physics, geophysics, metallurgy, and the physics of materials, with due attention to properties that were simply ignored at the usual scale of exploitation of continuum mechanics, for example, in hydraulics, hydrodynamics, the mechanics of structural members (the badly named "strength of materials"), and the mechanics of material bodies with oversimplified properties. The *mechanics of materials*, as understood nowadays, endeavors to examine materials at a finer scale than usual. The dynamics then draws a specific interest in the involved scale with typical lengths and frequencies characteristic of the materials under study. The dynamics often involves the notion of transport of mass (think of the flow of a fluid), but more generally it is some information that is transported in the form of a wave. In this context, we are mainly interested in materials of the deformable solid type. Such media, considered as continua – although some discreteness starts to be felt – are sometimes said to flow in an uncontrolled manner (think of the flow in plasticity) or a more controlled manner (viscoplastic bodies). However, while linear isotropic homogeneous elasticity remains the paragon against which all other behaviors are contrasted, the emerging *dynamics of materials* has incorporated such features as nonlinearity, inhomogeneity, and also some multiphysics with coupling with additional degrees of freedom and other properties than mechanical. With this, one enters the

The Complexity of Dynamical Systems. Edited by J. Dubbeldam, K. Green, and D. Lenstra
Copyright © 2011 WILEY-VCH Verlag GmbH & Co. KGaA, Weinheim
ISBN: 978-3-527-40931-0

framework of *generalized continua* [1]. This is needed due to the new ways of exploitation of materials and the existence of newly designed materials in which this complexity plays a fundamental role. It is in the light of these recent developments that we present a somewhat general approach illustrated mostly by one-dimensional (in space) examples. The dynamical properties exhibited are nonlinearity, dispersion, and couplings between wave modes. Even of greater interest are the possibilities of cooperation and competition between these different properties, yielding a very large spectrum of potentialities that will hopefully find uses in new devices.

Section 9.2 introduces the theory of elasticity as it is now commonly presented in modern continuum mechanics, first in the invariant frame of finite strain theory and then in the more familiar form of small strain elasticity. An original point here is the introduction of so-called conservation laws (related to Lie groups and Noether's invariant theory), in addition to the more standard balance of linear momentum, because this plays a particular role in wave studies. However, it also shows the close nature of some wave processes in elastic crystals with the wave mechanics of quantum fame. Of course, the local balance of linear momentum is the equation that provides the ubiquitous wave equation (d'Alembert's celebrated equation), but the introduction of physical nonlinearity (i.e., nonquadratic potential energies) and of a so-called weak nonlocality (technically equivalent to discreteness) proves to be essential in the contemporary focused interest in materials dynamics at a small scale. With the notion of strain gradient and that of hyperstress, we are thus inevitably led to dealing with a class of partial differential equations that yield a propagative but simultaneously dispersive character of related wave solutions. This is referred to here as the Boussinesq paradigm (Section 9.3) with its introduction of the notion of solitary waves and solitons. This is expanded at some length in Section 9.3 with examples from various systems and the appropriate numerical illustrations in 1D spatial systems but with one or two degrees of freedom. The notion of quasi-particles, born with those of photons and phonons, then enters the stage. Although some examples have become classics in that field (Korteweg–de Vries equations, sine-Gordon equation), other more realistic, but more complex, models are also introduced. As a matter of fact, Section 9.4 dealing with the fundamental problem of the propagation of phase transition fronts – by which dynamics a specimen becomes more in one phase than in another one on the application of some loading (e.g., mechanical impact, temperature) – shows how some of these complicated models appear quite naturally and can be interpreted in different fashions depending on the interest, formation, and idiosyncrasies of the concerned scientist. This is also duly illustrated and some analogies with some problems of mechanobiology are emphasized, nonetheless keeping in mind essential differences between inert matter and living biological tissues. Finally, Section 9.5 gives an overview of the so-called dynamic materials in which space- and time-varying properties can be duly exploited. This points the way to future developments in which we include the dynamics of fractal material sets.

9.2
Essentials of Elasticity

9.2.1
Finite Strain Elasticity

The elastic motion of a material particle of material coordinates **X** is given at time t by the following diffeomorphism (Figure 9.1):

$$\mathbf{x} = \bar{\mathbf{x}}(\mathbf{X}, t), \tag{9.1}$$

between the chosen reference configuration K_R and the actual configuration K_t in $E^3 \cong R^3$.

The physical velocity of **X** and the deformation gradient are defined by

$$\mathbf{v} := \left.\frac{\partial \bar{\mathbf{x}}}{\partial t}\right|_X, \quad \mathbf{F} := \left.\frac{\partial \bar{\mathbf{x}}}{\partial \mathbf{X}}\right|_t \equiv \nabla_R \bar{\mathbf{x}}. \tag{9.2}$$

For a first-order gradient theory of elasticity, the Lagrangian density L per unit volume (no dissipation) in the reference configuration of the continuous body reads

$$L = \bar{L}(\mathbf{v}, \mathbf{F}; \mathbf{X}, t) = K(\mathbf{v}; \mathbf{X}, t) - W(\mathbf{F}; \mathbf{X}, t), \tag{9.3}$$

where W is the elasticity potential and K is the kinetic energy given in a general manner by

$$K = \frac{1}{2}\varrho_0(\mathbf{X}, t)\mathbf{v}^2. \tag{9.4}$$

Accordingly, the *balance of linear physical momentum* (three components of the field **x** in physical space) reads in the absence of external body force

$$\left.\frac{\partial \mathbf{p}}{\partial t}\right|_X - \mathrm{div}_R \mathbf{T} = \mathbf{0}, \tag{9.5}$$

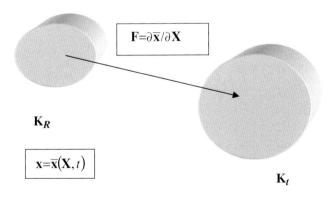

Figure 9.1 Finite deformation of a continuum.

wherein

$$\mathbf{p} := \varrho_0 \mathbf{v}, \tag{9.6a}$$

$$\mathbf{T} := \frac{\partial W}{\partial \mathbf{F}} \tag{9.6b}$$

are the (physical) linear momentum per unit reference volume and the first Piola–Kirchhoff stress. The *canonical conservation equations* [2] of energy and momentum are then given by

$$\left.\frac{\partial H}{\partial t}\right|_X - \nabla_R \cdot \mathbf{Q} = h \tag{9.7}$$

and

$$\left.\frac{\partial \mathbf{P}}{\partial t}\right|_X - \operatorname{div}_R \mathbf{b} = \mathbf{f}^{inh} \tag{9.8}$$

in which we have defined the following quantities:

$$H = K + W, \quad \mathbf{Q} = \mathbf{T} \cdot \mathbf{v}, \tag{9.9}$$

$$\mathbf{P} := -\varrho_0 \mathbf{v} \cdot \mathbf{F}, \tag{9.10a}$$

$$\mathbf{b} = -(L\mathbf{1}_R + \mathbf{T} \cdot \mathbf{F}), \tag{9.10b}$$

$$h := -(K/\varrho_0)\left.\frac{\partial \varrho_0}{\partial t}\right|_X + \left.\frac{\partial W}{\partial t}\right|_{expl}, \tag{9.11}$$

$$\mathbf{f}^{inh} := (K/\varrho_0)\left.\frac{\partial \varrho_0}{\partial \mathbf{X}}\right|_t - \left.\frac{\partial W}{\partial \mathbf{X}}\right|_{expl} \tag{9.12}$$

that are identified as the energy (Hamiltonian) density per unit reference volume, the material energy flux, the so-called canonical or material momentum, the Eshelby material stress tensor, the source of energy, and the material inhomogeneity force. More than often the symmetric material second Piola–Kirchhoff stress – or energetic stress – $\mathbf{S}$ is introduced in such a way that W assumes the following "objective" functional dependence:

$$W = \tilde{W}(\mathbf{E}; \mathbf{X}, \tilde{t}), \quad \mathbf{S} = \frac{\partial \tilde{W}}{\partial \mathbf{E}} = \mathbf{T}\mathbf{F}^{-T}, \tag{9.13}$$

where $\tilde{t} = t - t_0$ is a relative time and

$$\mathbf{E} = \frac{1}{2}(\mathbf{C} - \mathbf{1}_R), \quad \mathbf{C} := \mathbf{F}^T \mathbf{F}. \tag{9.14}$$

Equation 9.5 is to be solved under given boundary conditions and initial conditions that we skip here. Then the "identities" (9.7) and (9.8) are to be checked for the faithfulness of the obtained solution. This procedure has yielded an improvement of numerical methods (in finite difference and finite element methods; see, for example, Ref. [3]).

The following special cases are of obvious physical interest. *Materially inhomogeneous elastic materials* are elastic bodies in which both the reference density ϱ_0 and/or energy W depend explicitly on the material point $\mathbf{X}$. Material bodies in which density and/or energy depend explicitly on time are called *rheonomic* materials. Otherwise, they are said to be *scleronomic*. Materials in which only the reference density depends explicitly on time are said to be in a state of volumetric growth or resorption (such as in the growth of soft tissues). Materials in which only the energy W depends explicitly on the relative time $\bar{t}$ are said to be time evolving, and more particularly aging (such as in creep) or rejuvenating (clearly not frequent!). The last two properties are related to dissipation (nonzero source term on the right-hand side of Equation 9.7), in which case the second law of thermodynamics should be brought into the picture. In contrast, material inhomogeneity is not related to dissipation. As a conclusion to this point, we note that hyperelasticity is clearly a paradigmatic theory of fields. Now, the dynamic elasticity encapsulated in a Lagrangian such as (9.3) may be fully *nonlinear* as regards deformation processes, but it does not contain any characteristic length, except for inhomogeneous materials where the spatial variation of the material properties (density, elasticity coefficients) may provide such a length. Homogeneous elasticity exhibiting a characteristic intrinsic length of necessity involves gradients of higher order than the first, for example, the second material gradient of the placement or, just the same, the material gradient of the deformation gradient. That is, we may have to consider a generalization of (9.3) of the type

$$L = \bar{L}(\mathbf{v}, \mathbf{F}, \nabla_R \mathbf{F}; \mathbf{X}, t) = K(\mathbf{v}; \mathbf{X}, t) - W(\mathbf{F}, \nabla_R \mathbf{F}; \mathbf{X}, t), \tag{9.15}$$

where the inertial contribution is left unchanged. In this case, more lengthy representations than (9.6b) and (9.10b) have to be envisaged (see Ref. [4] for these developments).

It is only in scleronomic materials that matter density is strictly conserved, that is, in the present Piola–Kirchhoff format,

$$\left. \frac{\partial \varrho_0}{\partial t} \right|_X = 0. \tag{9.16}$$

9.2.2
Small Strain Elasticity

Following standard treatises, we note first that $\mathbf{u} = \mathbf{x} - \mathbf{X}$ is the elastic displacement and $\mathbf{e} = (\nabla \mathbf{u})_S$ denotes the small strain. We no longer distinguish between material and actual coordinates and call σ the symmetric Cauchy stress (in the absence of applied external couples). All the above equations are easily translated to this case, which does not mean loss of all nonlinearity but only of the so-called geometrical nonlinearity, physical nonlinearities remaining potentially present and of obvious interest in this contribution. In particular, we have thus the local balance of *physical (linear) momentum*:

$$\frac{\partial}{\partial t}\left(\varrho_0(\mathbf{x},t)\frac{\partial \mathbf{u}}{\partial t}\right) - \text{div}\,\sigma = \mathbf{0}, \quad \sigma = \frac{\partial \tilde{W}(\mathbf{e};\mathbf{x},\tilde{t})}{\partial \mathbf{e}}, \quad \mathbf{e} := (\nabla \mathbf{u})_S, \quad (9.17)$$

The balance of material momentum, on account of (9.17), yields the balance of *field momentum*. That is,

$$\frac{\partial}{\partial t}\mathbf{P}^f - \text{div}\,\mathbf{b}^f = \mathbf{f}^{\text{inh}}, \quad (9.18)$$

with

$$\mathbf{P}^f = -\varrho_0(\nabla\mathbf{u}) \cdot \frac{\partial \mathbf{u}}{\partial t}, \quad (9.19)$$

$$\mathbf{b}^f = -\left(L\mathbf{1} + \sigma \cdot (\nabla\mathbf{u})^T\right), \quad (9.20)$$

$$L = \frac{1}{2}\varrho_0\left(\frac{\partial \mathbf{u}}{\partial t}\right)^2 - \tilde{W}(\mathbf{e};\mathbf{x},\tilde{t}), \quad \mathbf{f}^{\text{inh}} = \frac{\partial L}{\partial \mathbf{x}}\bigg|_{\text{expl}}, \quad (9.21)$$

where $\tilde{W}$ need not be restricted to a quadratic function in $\mathbf{e}$. It may be of higher order than quadratic, even nonconvex. It is in this framework that $\mathbf{P}^f$ is also referred to as the *crystal momentum* for an elastic crystal after the pioneering work of Brenig [5]. As a matter of fact, for quantified elastic waves under the name of *phonons* (these are special types of so-called *quasi-particles* that are quantum mechanically associated with definite types, here elastic, of vibrations), this is the momentum that directly quantifies to *de Broglie's formula*

$$\mathbf{P}^f = \hbar\mathbf{K}, \quad (9.22)$$

where $\hbar$ is Planck's reduced constant and $\mathbf{K}$ is the wave vector. Note that in the nonlinear framework, the wave vector is a *covariant vector* (since it is the dual of material position $\mathbf{X}$), and thus we understand the proportionality relation between the naturally covariant vector $\mathbf{P}$ or $\mathbf{P}^f$ and $\mathbf{K}$ (see pp. 35–37 and Chapter 9 of Ref. [4]). Because of its role in *wave propagation*, $\mathbf{P}^f$ is also referred to as the *wave momentum*.

9.2.3
One-Dimensional Motion

For a one-dimensional motion (along coordinate x), all quantities become scalars; but general conclusions hastily drawn from this very special case may be dangerous or definitely wrong in the three-dimensional case. In this framework, we note that $\mathbf{e}$ yields $e = u_x := \partial u/\partial x$ and $\mathbf{v}$ yields $v = u_t := \partial u/\partial t$. If we consider the case of linear isotropic homogeneous elasticity (for sclenomic materials) in one dimension of space with constant Hooke's modulus E and characteristic wave speed

9.2 Essentials of Elasticity

$c_0 = (E/\varrho_0)^{1/2}$, the balance of linear physical momentum yields the ubiquitous wave equation

$$u_{tt} - c_0^2 u_{xx} = 0, \tag{9.23}$$

in an obvious notation. The corresponding energy equation, obtained by multiplying (9.23) by u_t, reads

$$\frac{\partial}{\partial t}\left(\frac{1}{2}\varrho_0 u_t^2 + \frac{1}{2}E u_x^2\right) - \frac{\partial}{\partial x}\left((E u_x) u_t\right) = 0. \tag{9.24}$$

From its very definition, the balance of canonical (here field) momentum, obtained by multiplying (9.23) by u_x, reads

$$\frac{\partial}{\partial t}(-(\varrho_0 u_t) u_x) - \frac{\partial}{\partial x}\left(\left(\frac{1}{2}E u_x^2 - \frac{1}{2}\varrho_0 u_t^2\right) - (E u_x) u_x\right) = 0. \tag{9.25}$$

The last two equations can also be written in the following remarkable form:

$$\frac{\partial H}{\partial t} + c_0^2 \frac{\partial P^f}{\partial x} = 0, \quad \frac{\partial P^f}{\partial t} + \frac{\partial H}{\partial x} = 0. \tag{9.26}$$

Even more remarkably, this shows that both H and P^f satisfy the original wave equation(9.23) since by elimination between the two equations (9.26) one obtains

$$H_{tt} - c_0^2 H_{xx} = 0, \quad (P^f)_{tt} - c_0^2 (P^f)_{xx} = 0. \tag{9.27}$$

Interesting as it is, this result is however misleading for it is an artifact of the one-dimensional formulation. In effect, the energy equation is normally a scalar one, while the balance of field momentum is covectorial. The misleading symmetry induced by the one-dimensional nature between these two equations was noted by W.D. Hayes ([6], pp. 23–24) when he wrote down equations (9.26) as two quadratic invariant equations deduced from (9.23) (without the present Eshelbian framework and the consequences (9.27)). Hayes simply comments that the "freedom of generating new solutions by differentiation or integration must be kept in mind, as these generate new conservation laws." This is what happens in the theory of solitonic structures.

9.2.4
Physical Nonlinearity

In order to avoid combining various complexities, we consider here materially homogeneous scleronomic materials. If the energy W is expanded in power series of $e = u_x$ for one-dimensional motions, the equation of balance of linear momentum will yield the *nonlinear wave equation*

$$u_{tt} - c^2(u_x)_{xx} = 0 \tag{9.28a}$$

$$c^2(u_x) = c_0^2\left(1 + a_1 e + a_2 e^2 + \cdots\right) \tag{9.28b}$$

where the coefficients $a_i, i = 1, 2, \ldots$, are constructed from the elasticity coefficients of various orders [7]. The a_i's are finite (combination of material coefficients) in magnitude and the small parameter here is the solution itself, e. An ε can be introduced only as a marker of the smallness of the strain. This suggests a study of equations such as (9.28a) by applying various techniques of perturbation. Note that (9.28a) can also be cast as a system of first-order nonlinear partial differential equations for the couple of functions (e, v):

$$v_t = c^2(e)e_x, \quad e_t = v_x, \tag{9.29}$$

where the second equation expresses compatibility between second-order space–time derivatives. Exact equations (9.29) are typically treated within the theory of *quasi-linear hyperbolic systems*. Such systems exhibit simple wave solutions and are prone to developing shock wave formation after a certain breaking distance (see Ref. [7], Chapter 1). This is a standard subject matter in textbooks (theory of linear and nonlinear characteristics). In the case of a very weak nonlinearity, imputing sinusoidal signals and using perturbation schemes for distances of propagation much less than the mentioned breaking distance, one may exhibit such effects as the production of harmonics and the phenomenon of anisochronism. Coupling with quasi-static electric fields in (appropriately noncentrosymmetric anisotropic) piezoelectrics enhances, and allows the control and application of, these phenomena in signal processing [7].

What developed in the 1960s–1970s was a more challenging perspective where the effects of *nonlinearity* may be balanced by those of *dispersion*, giving rise to the notions of *solitary waves* and *solitons*.

9.3
The Boussinesq Paradigm and Akin Nonlinear Dispersive Systems

9.3.1
The Standard Boussinesq (BO) Equation in Elastic Crystals

An often cited exemplary equation in the present context is the so-called Boussinesq equation that has both fluid mechanics and crystal mechanics origins (cf. [8]). In appropriate units, this may be considered as the field equation of elasticity deduced from a second gradient of displacement elasticity or strain gradient elasticity (see the Lagrangian (9.15). It is a one-dimensional (in space) dispersive but nondissipative nonlinear model deduced from the following Lagrangian and strain energy densities:

$$L = \frac{1}{2} u_t^2 - W(u_x, u_{xx}), \quad W = \frac{1}{2}\left(u_x^2 + \frac{2}{3}\varepsilon u_x^3 + \varepsilon \delta^2 u_{xx}^2\right). \tag{9.30}$$

Here the system is materially homogeneous and scleronomic. The characteristic (linear) wave speed c_0 has been set equal to 1. The resulting Euler–Lagrangian equation is the following nonlinear dispersive wave equation (the celebrated Boussinesq equation):

$$u_{tt} - u_{xx}(1 + \varepsilon u_x) - \varepsilon \delta^2 u_{xxxx} = 0, \tag{9.31}$$

where ε is an infinitesimally small parameter characteristic of the *nonlinearity* and δ is a length characteristic of a weak *nonlocality* (dispersion) of the modeling. This equation, where both nonlinearity and dispersion intervene at the same order, is none other than the field equation (balance of linear physical momentum)

$$\frac{\partial}{\partial t}(u_t) - \frac{\partial}{\partial x}(\sigma^{\text{eff}}) = 0, \tag{9.32}$$

with

$$\sigma^{\text{eff}} = \frac{\delta W}{\delta u_x} = \sigma - m_x, \quad \sigma = \frac{\partial W}{\partial u_x}, \quad m = \frac{\partial W}{\partial u_{xx}}, \tag{9.33}$$

where m is the scalar degenerate form of the *hyperstress* tensor. The corresponding energy equation is easily established, while the balance of field momentum is given by

$$\frac{\partial}{\partial t}(P^{\text{f}}) - \frac{\partial}{\partial x}(b^{\text{eff}}) = 0, \tag{9.34}$$

with

$$P^{\text{f}} = -u_x u_t, \quad b^{\text{eff}} = -(L + u_x \sigma + 2u_{xx} m) - (u_x m)_x. \tag{9.35}$$

Integrating this over the whole real line R, we obtain the global conservation of field momentum as

$$\frac{dP(R)}{dt} = [b^{\text{eff}}]_{-\infty}^{+\infty}, \tag{9.36a}$$

$$P(R) := \int_R P^{\text{f}} dx. \tag{9.36b}$$

This is a Newtonian-like equation of motion in which the jump between the two end values of the effective Eshelby "tensor," here reduced to a scalar, plays the role of a *driving force*. A material inhomogeneity or an additional externally prescribed term on the right-hand side of (9.32) will also bring additional driving forces on the right-hand side of (9.36a).

It happens that the Boussinesq equation (9.31), like many equations belonging to the same class (what we called the Boussinesq paradigm; cf. [8]), possesses strongly localized nonlinear solutions (kinks for u and humps for u_t) with appropriate space and time derivatives vanishing at infinities. Thus, Equation 9.36 reduces to the equation of an *inertial motion* for these solutions:

$$\frac{dP(R)}{dt} = 0 \quad \text{and} \quad \frac{dH(R)}{dt} = 0, \quad H(R) := \int_R H dx. \tag{9.37}$$

If an external force perturbing density $\mu f(x)$ were acting on the right-hand side of (9.32), μ being a small parameter, then the first of (9.37) would be perturbed in the following way:

$$\frac{dP}{dt} = -\varepsilon^2 \int_R u_x f(x) dx, \qquad (9.38)$$

where we took $\mu = 0(\varepsilon)$, and one has to find from (9.38) the modulation (due to the perturbation) of the parameters of the localized nonlinear wave.

The Boussinesq equation has acquired much of its celebrity through the derived equation called the *Korteweg–de Vries* (KdV) *equation*, in fact the one-directional wave equation deduced from (9.31) by means of the so-called *reductive perturbation method* (cf. [9]). After appropriate nondimensionalization, this equation reads as follows:

$$\frac{\partial v}{\partial t} + v \frac{\partial v}{\partial x} + \beta \frac{\partial^3 v}{\partial x^3} = 0, \quad \beta = \frac{c_0^2}{2k_0^2} > 0, \qquad (9.39)$$

where c_0 and k_0 are characteristic speed and wave number, respectively. This equation (Korteweg, de Vries, Boussinesq, and Rayleigh) admits *exact solitary waves* solutions in the form of a *hump*, $u(x \to \pm\infty) = 0$, of the type

$$v = v_0 \operatorname{sech}^2 \left(\frac{x - ct}{\Delta} \right), \qquad (9.40)$$

under the condition that speed c and amplitude v_0 be related by

$$v_0 = 3c, \quad \Delta^2 = 2c_0/ck_0^2; \quad \text{hence,} \quad v_0 \Delta^2 = 6c_0/k_0^2 = \text{const.} \qquad (9.41)$$

Accordingly, the faster the wave, the narrower its profile. We also note that the wavelike solution is *supersonic* as $c > c_0$. We also remark that a simple wave trial solution of the type $u \approx \bar{u}(x - ct)$ in (9.31) would have resulted in a kink-like solution in the form of a *tanh* solution (of which (9.40) is a derivative). The existence of such solutions is due to a strict compensation between nonlinear effects (steepening of a solution to form a shock; signal with higher amplitude would go faster) and dispersive ones (broadening of the signal due to differing speeds of propagation for various Fourier components).

Equation (9.39) admits exact localized solutions such as (9.40), but such solutions, although truly nonlinear, in some sense practically superimpose linearly each other since two such solutions traveling in opposite direction interact without further perturbation than a change in phase, recovering their individuality after encounter (collision). This is the property of being *solitonic per se* in a strict mathematical sense. In nonexactly solitonic systems, the interactions of "individuals" are usually accompanied by the production of radiation. Figure 9.2 shows the case of an encounter of almost solitonic structures (little amount of radiation in part (a)) for a slightly perturbed exact solitonic Boussinesq system exhibiting two critical speeds. Practically no radiation is exhibited for so-called subcritical solitons in this system (b). As a matter of fact, it was soon realized by the pioneers (and creators) of soliton theory, for example, Kruskal and Zabusky [10] and Kruskal [11], for the KdV equation, that systems of equations prone to the pure solitonic type of dynamic behavior admit new conservation laws in addition to the usual ones (the latter are the traditional balance

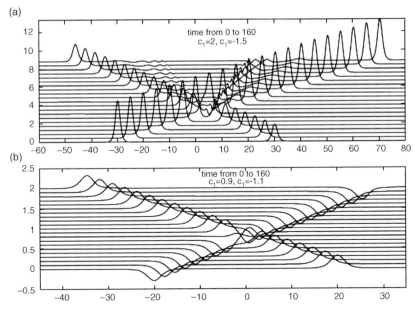

Figure 9.2 Two counterpropagating "hump" solitons (a) or antisolitons (b) exhibiting radiation in the course of interaction; horizontally: space; vertically: increasing time slices.

laws). It was further shown that for exactly integrable systems (those admitting indeed true soliton solutions), there exists an infinity of such conservation laws and special algorithms were developed to generate these laws (in this regard, cf. [12], Section 9.1.6, and [13], Chapter 5). If the field equations used to describe solitons are derived in a field theoretic context from a Lagrangian (or Hamiltonian), then these new conservation laws correspond to *symmetry properties* and result from the application of Noether's theorem [14]. However, only a few of these conservation laws can bear an easily grasped physical significance. The reader may now have realized where we want to lead him since those easily meaningfully interpreted conservation laws are those that pertain to our Eshelbian or canonical framework. These conservations laws are those that are more critically related to the particle-like features of true solitons – which are thus kinds of *quasi-particles* – in the course of so-called elastic collision or interactions. Canonical *momentum* is one of these features. Other such features are *mass* and *energy*, these three quantities forming if possible a true *point mechanics* of which the type will depend on the starting system of partial differential equations. Newtonian and Lorentzian–Einsteinian point mechanics are examples of such mechanics. Others can be created that no direct evidence could bring to the fore [15].

Of course, a duality between soliton-like solutions of some systems issued from quantum physics and elementary particles was rapidly established by nuclear and high-energy physicists [16]. We must also remember the attempts of L. de Broglie and D. Bohm to reconcile quantum physics and a causal interpretation by introducing, in a nonlinear framework, the notion of *pilot wave* guiding the amplitude of the probability ($|\psi|^2$) of presence of a particle as a wave of singularity for which conservation laws and a hydrodynamic analogy play an essential role ([17], pp. 113–124). It is possible that the present developments bear some relationship to this, but we emphasize that we are mostly interested in macroscopic problems issued from engineering sciences and phenomenological physics (e.g., in the crystalline state) although at a smaller scale than usual in traditional engineering.

9.3.2
The "Good" Boussinesq Equation

This is a nonlinear dispersive wave equation of the form

$$u_{tt} - u_{xx} - (u^2 - u_{xx})_{xx} = 0. \tag{9.42}$$

The essential difference from (9.31) lies in the change in sign in front of the fourth-order space derivative. The reason for that is the bad linear dispersive behavior exhibited by the original equation (9.31) (the so-called *anomalous dispersion*). Equation 9.42 also corresponds to a one-dimensional model of strain gradient elasticity. Upon introducing auxiliary variables q and w, this can be rewritten as the following Hamiltonian system [18]:

$$u_t = q_x, \quad w = u_x, \quad q_t = w_x^2 + w_{xx} - w, \tag{9.43}$$

in which the first two are mere definitions of q and w. The mass M, momentum P, and energy E of soliton solutions of (9.42) or (9.43) are given by

$$M = \int_R u \, dx, \tag{9.44}$$

$$P(R) = -\int_R uq \, dx, \tag{9.45}$$

and

$$E(R) = \frac{1}{2} \int_R \left(q^2 + w^2 + u^2 + \frac{2}{3} u^3 \right) dx. \tag{9.46}$$

As the system considered is exactly integrable, the quantities just defined are strictly conserved. But their expressions may look somewhat awkward. However, introducing the potential $\bar{u}$ by $u = \bar{u}_x$ with the condition $\bar{u}(x \to -\infty) = 0$, it is verified that

$$M = [\bar{u}]_{-\infty}^{+\infty}, \quad \bar{u}_t = q, \quad uq = \bar{u}_x \bar{u}_t, \quad \frac{1}{2} q^2 = \frac{1}{2} \bar{u}_t^2, \tag{9.47}$$

so that M has the same interpretation as in the KdV case, while P and E indeed take their canonical definitions in terms of the potential $\bar{u}$. Simultaneously, in terms of elasticity theory, it is $\bar{u}$ that has the meaning of a displacement, while u is a strain *per se*. But accepting the general philosophy of continuum mechanics, we can also consider (9.42) as a field equation issued from second-grade nonlinear elasticity and multiply it by u_x and integrate by parts to arrive at the equation of field momentum (cf. Equation 9.34):

$$\frac{\partial}{\partial t} p^f - \frac{\partial}{\partial x} b^{\text{eff}} = 0, \quad p^f := -u_x u_t. \tag{9.48}$$

9.3.3
Generalized Boussinesq Equation

There are several ways to generalize the BO equation, obviously leading to nonexactly integrable systems. One such system is obtained while studying the ferroelastic phase transition as a dynamical process in elastic crystals [19]. Here, we examine the generalization of this modeling proposed by Christov and Maugin [20] when approaching the difference scheme of lattice dynamics in a more accurate way than usually done. With $s = v_x$, a *shear strain*, from a lattice dynamics approach and a long-wavelength limit while neglecting coupling with other strain components, one obtains the following type of equation:

$$s_{tt} - c_T^2 s_{xx} - (F(s) - \beta s_{xx} + s_{xxxx})_{xx} = 0, \tag{9.49}$$

where F is a polynomial in s starting with second degree (e.g., a nonconvex function admitting three minima), c_T is a characteristic speed, and β is a positive scalar. It can be said that both the nonlinearity and dispersion have been increased compared to the classical BO equation [21]. Equation 9.49 is *stiff* in the sense that it involves a sixth-order space derivative, a situation that obviously imposes rather strong limit conditions at infinity or at the ends of a finite interval in numerical simulations. In spite of its apparent complexity, (9.49) admits solitary wave solutions [20], which involve the ubiquitous *sech* function (but at the fourth power) for a single value of the phase speed – the existence of different solitary wave solutions with a continuous spectrum for c was shown numerically [22]. But it is true that for a velocity too close to c_T, these solutions are not able to preserve their shape and eventually they transform into pulses that, in turn, exhibit a self-similar (kind of "big bang") behavior as long as the amplitude of the pulse decreases while its support increases (a phenomenon analogous to a *redshift*). These pulses practically pass through each other without changing qualitatively their shapes – save the redshifting – with perfect conservation of "mass" and "energy," so that these pulses may qualitatively be claimed to be "solitons" [20]. We refer to Ref. [23] for the point mechanics associated with the system (9.49). A strongly implicit conservative finite difference scheme must be used in numerical simulations in order to always preserve both mass M and energy E. Figure 9.3 illustrates the counterpropagation of two identical so-called "Kawahara solitons" of an appropriate speed in system (9.49). Some radiation is exhibited for this nonexactly integrable system, but there is no sensible phase shift after the encounter.

So far we considered mechanical systems with only one degree of freedom, even though all in kinds of generalized elasticity theory. Because of the additive definition of the canonical entities (see the canonical definitions in Ref. [4]), the

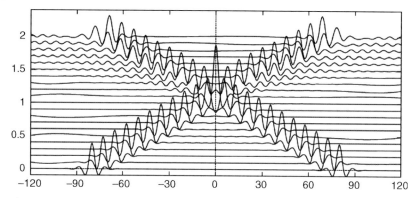

Figure 9.3 Two counterpropagating identical "Kawahara" solitons (not exactly integrable system exhibiting little amount of radiation but no sensible phase shift after interaction).

consideration of several degrees of freedom in pure mechanics or in a coupled field theory is rather simple.

9.3.4
Mechanical System with Two Degrees of Freedom

When the small coupling between the v degree of freedom of the previous example and the longitudinal displacement u and elongation strain $e = u_x$ is kept [24]– but remember that the ferroelastic phase transition is driven through s, e being only a secondary subsystem – we have the following coupled wave system:

$$s_{tt} - c_T^2 s_{xx} + \left(s^3 - s^5 + 2\gamma\, se + \alpha\, s_{xx}\right)_{xx} = 0, \tag{9.50a}$$

$$e_{tt} - c_L^2 e_{xx} + \gamma\left(s^2\right)_{xx} = 0, \tag{9.50b}$$

$$s = v_x, \quad e = u_x, \quad (v, u) \in R^2, \tag{9.50c}$$

where γ is a coupling coefficient and c_L is a second characteristic speed larger than c_T. This system looks formidable. Still it admits exact analytical solitary wave solutions that do represent the various transitions possible between austenite and two martensitic variants of opposite shear. The sixth-order space derivatives have been discarded. With the quadruplet (s, q, e, r), we can rewrite system (9.50) as the following Hamiltonian system:

$$s = q_{xx}, \tag{9.51a}$$

$$e_t = c_L r_x, \tag{9.51b}$$

$$q = c_T^2 s - s^3 + s^5 - 2\gamma se - \alpha s_{xx}, \tag{9.51c}$$

$$r_t = c_L e_x - (\gamma/c_L)(s^2)_x. \tag{9.51d}$$

The associated total mass, momentum, and energy are given by

$$M = \int_R s\,dx = [v]_{-\infty}^{+\infty}, \tag{9.52}$$

$$P(R) = \int_R (sq_x + c_L e_r)\,dx, \tag{9.53}$$

$$E(R) = \frac{1}{2}\int_R \left(\left(q_x^2 + c_L^2 r_x^2\right) + \left(c_T^2 s^2 + c_L^2 e^2\right) - 2\gamma\, es^2 - \frac{1}{2}s^4 + \frac{1}{3}s^6 + \alpha(s_x)^2 \right) dx. \tag{9.54}$$

With definition (9.53), which we let the reader show to be true in agreement with the canonical definition, and *ad hoc* conditions at infinity, we obtain a global balance of field momentum in the inhomogeneous form:

$$\frac{d}{dt}P(R) = \bar{F} := -\frac{1}{2}\left[\alpha s_x^2 + c_L^2 e^2\right]_{-\infty}^{+\infty}. \tag{9.55}$$

This shows what conditions should apply to have conservation of global field momentum strictly enforced so as to have an *inertial motion* of the possible shapes exhibited by the considered system if we work on a necessarily bounded finite interval in a numerical simulation [22].

9.3.5
Sine-Gordon Equations and Associated Systems of Equations

9.3.5.1 Pure Sine-Gordon System

The sine-Gordon equation is the well-known one-dimensional (in space) partial differential equation:

$$u_{tt} - u_{xx} + \sin u = 0, \tag{9.56}$$

where both nonlinearity and dispersion are contained in the *sin* function. This ubiquitous equation that can also be written as *Enneper's equation* of surface geometry as

$$u_{\xi\zeta} - \sin u = 0 \tag{9.57}$$

by introducing right- and left-running characteristic coordinates ξ and $\zeta = x \pm t$ appears in many fields of physics, especially while studying the structure of magnetic domain walls and Josephson junctions in superconductivity [25]. From the mechanical viewpoint, such an equation can be obtained while studying the torsion of some bars [26] and, above all, as an elementary model of dislocation motion in the so-called Frankel–Kontorova model [27] – a linear atomic chain placed in a sinusoidal potential landscape. This remarkable equation is exactly integrable (i.e., admits true soliton solutions) and is Lorentz invariant. It admits single subsonic solitary wave solutions of the *kink* form. But contrary to previously examined cases, the amplitude of such solutions does *not* depend on the velocity. They are *topological* solitons. The point mechanics associated with the quasi-particles representing such solitons is Lorentzian.

Indeed, viewed as an *elastic* system, (9.56) is derivable from the following Lagrangian–Hamiltonian framework where the sinusoidal term should be interpreted as the action of an external source (the already mentioned periodic substrate of the Frenkel–Kontorova modeling) since the classical elastic energy cannot depend explicitly on u:

$$L = \frac{1}{2}u_t^2 - \frac{1}{2}u_x^2 - (1-\cos u), \quad p = \frac{\partial L}{\partial u_t}, \tag{9.58}$$

$$H = pu_t - L = \frac{1}{2}(p^2 + u_x^2) + 2\sin^2(u/2), \tag{9.59}$$

with Hamiltonian equations

$$u_t = \frac{\partial H}{\partial p} = p, \quad p_t = -\frac{\delta H}{\delta u} = u_{xx} - \sin u. \tag{9.60}$$

A kink (2π solution in u) or an antikink (-2π solution) may be considered as a quasi-particle with rest mass M_0, momentum P, and energy E given by

$$M_0 = 8 = E(0), \tag{9.61}$$

$$P = P(R) = \int_R (-u_x u_t) dx = 8\gamma c = Mc = \frac{M_0 c}{(1-c^2)^{1/2}}, \tag{9.62}$$

$$E = E(R) = \int_R H dx = 8\gamma = E(c), \tag{9.63}$$

with the classical relationship between the triplet (M_0, P, E) typical of Lorentzian point mechanics (the characteristic speed of relativity, the light velocity in vacuum, here is one):

$$E^2(c) = M_0^2 + P^2(c), \tag{9.64}$$

while for Newtonian point mechanics we would have

$$E = P^2/2M_0, \quad P = M_0 c. \tag{9.65}$$

Equations 9.62 and 9.64 obey canonical definitions and the given estimates are based on the form of the following kink solution:

$$u(x,t) = \bar{u}(\xi) = 4\tan^{-1}(\exp \pm \gamma(\xi - \xi_0)), \tag{9.66}$$

wherein

$$\xi = x - ct, \quad \gamma = (1-c^2)^{-1/2}, \quad |c| < 1. \tag{9.67}$$

Of course, for a fixed c, the three quantities M, P, and E are strictly conserved, that is, for kink and antikink solutions; in particular, we have a standard *inertial* equation of motion for a relativistic quasi-particle:

$$\frac{dP}{dt} = \frac{d}{dt}\left(\frac{M_0 c}{\sqrt{1-c^2}}\right) = 0. \tag{9.68}$$

9.3.5.2 Sine-Gordon–d'Alembert Systems

While studying magnetic domain walls in elastic ferromagnets [28], these authors were led to introducing systems of the following type:

$$\phi_{tt} - \phi_{xx} - \sin\phi = \eta u_x \cos\phi, \quad u_{tt} - c_T^2 u_{xx} = -\eta(\sin\phi)_x. \tag{9.69}$$

Here ϕ is twice the angle of rotation (of magnetic spins in a plane parallel to the x-axis of propagation and the polarization of the transverse elastic displacement u; the so-called *Néel wall* in ferromagnetism), c_T is a characteristic (transverse) elastic speed, and η is representative of a magnetostrictive magnetomechanical coupling. A similar problem arises in the ferroelectricity of deformable crystals of the polar type where η then is related to electrostriction in electroelasticity [29]. The system (9.69) couples linearly a sine-Gordon equation and a linear wave equation, and thus deserves its name (coined by Kivshar and Malomed [30]). Accordingly, this system is *not* exactly integrable from the point of view of soliton theory because the u-subsystem induces *radiations* during soliton interactions [31], but exact one-soliton solutions are known to exist analytically [29].

Ignoring the physical origin of the function ϕ, we may consider (9.69) as a two-degrees-of-freedom nonlinear elastic dispersive system with displacement components u and ϕ, where an external sinusoidal force acts only on the ϕ degree of freedom. We can then exploit the standard canonical formalism [4], while noting the additive property over field components of the canonical expressions. We thus write

$$L = \frac{1}{2}(u_t^2 + \phi_t^2) - W(u_x, \phi_x, \phi), \tag{9.70}$$

$$W = \frac{1}{2}(\phi_x^2 + c_T^2 u_x^2) + \eta u_x \sin\phi - (1 + \cos\phi), \tag{9.71}$$

$$H = p_u u_t + p_\phi \phi_t - L, \quad p_u = u_t, \quad p_\phi = \phi_t, \tag{9.72}$$

$$\sigma = \frac{\partial W}{\partial u_x}, \quad \mu = \frac{\partial W}{\partial \phi_x}, \tag{9.73}$$

and

$$p^f = -(u_x u_t + \phi_x \phi_t), \quad b = -(L + \sigma u_x + \mu \phi_x). \tag{9.74}$$

Therefore, the following local and global balances of field (wave) momentum hold, in the canonical form:

$$\frac{\partial P^f}{\partial t} - \frac{\partial b}{\partial x} = 0, \quad \frac{d}{dt} P(R) = [b]_{-\infty}^{+\infty}, \quad P(R) := \int_R P^f dx. \tag{9.75}$$

For solitary wave solutions for which all derivatives vanish at infinity, b vanishes at infinity and $P(R)$ is strictly constant for a fixed velocity. Exact solutions are given in Ref. [29] and the interactions of individuals (with accompanying radiation) are exhibited in Ref. [31]. Elastic systems with a kind of micropolar internal structure (microstructure of the rotational type) exhibit solitonic solutions with a similar dynamical behavior [32, 33]. The quasi-particle dynamics remains essentially Lorentzian but with perturbations due to the wave component u.

9.4
A Basic Problem of Materials Science: Phase Transition Front Propagation

9.4.1
Some General Words

A full understanding of the phenomenon of the propagation of phase transition fronts in deformable crystals – metals and alloys – is one of the essential problems of contemporary materials science and mechanics at both theoretical and application levels. This unique problem can be examined at three different scales: (i) *microscopic scale* (lattice dynamics) in the absence of thermodynamical irreversibility, (ii) *mesoscopic scale* (exploitation of continuum thermomechanical equations in a structured front), and (iii) *macroscopic scale*, that of engineering applications. The first scale inspired by the Landau–Ginzburg theory, although discrete to start with, deals with nonlinear localized waves (solitonic structures: solitary wave, soliton complexes) where nonlinearity and dispersion (discreteness) are the main ingredients. The elements recalled here deal with such wave dynamics. The second scale involves nonlinearity, dispersion, and dissipation (viscosity). The third scale is the one at which the front is seen as an irreversibly driven singular surface and where macroscopic thermodynamics (theory of irreversible processes) and numerical methods such as finite element and finite volume methods are used in conjunction with a criterion of progress.

The three scales are reconciled by the fact that all solutions satisfy the same *Hugoniot conditions* sufficiently far away from the front, whether structured as a solitonic or dissipative structure or viewed as a singular surface. This multilevel, multiphysics approach gathers the viewpoints of condensed matter physicists (microscale), applied mathematicians (mesoscale), and engineers (macroscale), and even that of the theoretical physicist via the inclusive notion of quasi-particles and the underlying and pervasive invariant theoretical framework. In all cases, the notion of *driving force* is involved being either set equal to zero or very active indeed.

9.4.2
Microscopic Condensed Matter Physics Approach: Solitonics

The first approach considered is that dealing with the *microscale* of lattice dynamics *in a perfect lattice*, so that there is no dissipation and effects of temperature are not involved, except perhaps in the phase transition parameter. This was particularly envisaged by Falk [34], Pouget [19], and Maugin and Cadet [24]. This approach allows one to readily obtain a dynamical representation of a phase boundary (here a kink) as a *solitonic structure* for a two-degrees-of-freedom, but essentially one-dimensional, system. In this context, the continuum model obtained in the *long-wavelength limit* is that of a *nonlinear elastic body with first gradients of strains* taken into account but no dissipation. This long-wave limit is admissible because the transition layer between two phases, although thin (perhaps a few lattice spacings), is nonetheless large enough. Numerical simulations can be performed *directly* on the lattice. The elastic potential is nonconvex in general. This is exemplified by considering a one-dimensional (x), two-degrees-of-freedom, lattice with transverse (main effect) and longitudinal (secondary effect) displacements from the initial position. In the so-called *long-wave limit* where the discrete dependent variables (strains) s_n and e_n vary slowly from one lattice site to the next and they can be expanded about the reference configuration (na, 0), the discrete equations yield a system of two (nondimensionalized) coupled partial – in (x, t) – differential equations (with an obvious notation for partial x and t derivatives), which is none other than the system of equations (9.50), where, we remind the reader, s and e are the shear and elongation strains, γ is a coupling coefficient, and α is a nonlocality parameter. Parameters c_T and c_L are the characteristic speeds of the linear elastic system. This corresponds to stresses and energy density given by

$$\sigma_S = \bar{\sigma}_S - m_x, \quad \sigma_e = \frac{\partial W}{\partial e}, \quad \bar{\sigma}_S = \frac{\partial W}{\partial s}, \quad m = \frac{\partial W}{\partial s_x} \tag{9.76}$$

and

$$W(s, e, s_x) = \frac{1}{2}\left(c_T^2 s^2 - \frac{1}{2}s^4 + \frac{1}{3}s^6 + c_L^2 e^2 - 2\gamma s^2 e + \alpha(s_x)^2\right). \tag{9.77}$$

In other words, Equations 9.50a and 9.50b are none other than the x-derivatives of the balance of (physical) linear momentum for a continuum made of a nonlinear, homogeneous elastic material with strain gradients – with both nonlinearity and strain gradients relating only to the *shear* deformation. As already noticed, this apparently complicated system still admits exact dynamical solutions of the *solitonic type*. A thorough discussion of the existence of solitary wave-like solutions – such solutions connecting two different or equivalent minimizers (i.e., two phases) of the potential energy – was given by Maugin and Cadet [24]. The remarkable fact is that such complicated solutions are shown (by computation) to satisfy the following (temperature-independent) *Hugoniot* condition between *states at infinity*:

$$\text{Hugo} := [\bar{W}(s, e \text{ fixed}) - \langle \bar{\sigma}_S \rangle s] = 0, \tag{9.78}$$

where $\bar{\sigma}_s$ is the shear strain *without* strain gradient effect and $\bar{W}$ is the elastic energy with such effects similarly neglected. Obviously, gradient effects play a significant role only within the rapid transition zone that the kink solution represents, while outside this zone the state is practically spatially uniform, although different on both sides of the localized front. Here, we have used the following definitions for the jump and mean value of any quantity *a*:

$$[a] := a(+\infty) - a(-\infty), \quad \langle a \rangle := \frac{1}{2}(a(+\infty) + a(-\infty)). \tag{9.79}$$

Equation 9.78 is typical of the *absence of dissipation* during the transition, in general a working hypothesis that is *not* realistic. Furthermore, it can in fact be rewritten as the celebrated *Maxwell's rule of equal areas*.

In the same conditions, the corresponding dynamic solitary wave-like solution satisfies the quasi-particle inertial motion:

$$\frac{d}{dt}P(R) = 0, \tag{9.80}$$

with vanishing driving force on the right-hand side, that is, Equation 9.55 with vanishing right-hand side because of the asymptotic behavior of the solutions.

9.4.3
Macroscopic Engineering Thermodynamic Approach

This is a local viewpoint that refers to the fact that it is assumed at each instant of time that the thermoelastic solution is known by any means – analytical, but more than often, numerical – on both sides of the singular surface Σ (cf. Figure 9.4) so that one can compute a *driving force* acting on Σ. Further progress of Σ must not contradict the second law of thermodynamics. The latter, therefore, governs the local evolution of Σ that is generally *dissipative*, although no microscopic details are made explicit to justify the proposed expressions. The approach is *thermodynamical* and *incremental* (in total analogy with modern plasticity). All physical mechanisms responsible for the phase transformation are contained in the *phenomenological macroscopic* relationship given by the *local criterion of progress* of Σ. Without entering details that can be found in

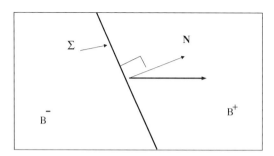

Figure 9.4 Phase transition front seen as a moving zero-thickness singular surface separating the two phases.

several studies [35–37] and considering from the outset the finite strain framework, we remind the reader that at any regular point in the body (i.e., on both sides of Σ), we have the balance of (physical) linear momentum and the future heat equation written in the Piola–Kirchhoff form for a *heat-conducting thermoelastic material* ($W(\mathbf{F},\theta)$ in general is different on both sides of Σ, and generally nonconvex in its first argument and concave in the second one – the thermodynamical temperature θ). But while each phase is materially homogeneous, the presence of Σ is a patent mark of a loss of translational symmetry on the overall body, hence the consideration of a global material inhomogeneity. The field equation capturing this breaking of symmetry is the jump relation associated with the equation of momentum *on the material manifold*, that is, what we have called the balance of material momentum in previous sections. This jump equation, together with that for entropy, governs the phase transition phenomenon at Σ. These equations a priori read

$$\mathbf{N} \cdot [\mathbf{b} + \bar{\mathbf{V}} \otimes \mathbf{P}] + \mathbf{f}_\Sigma = 0, \quad \mathbf{N} \cdot [\bar{\mathbf{V}}S - (\mathbf{Q}/\theta)] = \sigma_\Sigma \geq 0, \tag{9.81}$$

where the last inequality is a statement of the second law of thermodynamics at Σ and $\mathbf{N}$ is the unit normal to Σ oriented from the *minus* to the *plus* side, and we defined the jumps and mean values at Σ by (cf. Equation 9.79)

$$[a] := a^+ - a^-, \quad \langle a \rangle := \frac{1}{2}(a^+ + a^-), \tag{9.82}$$

where $a^\pm$ are the uniform limits of a in approaching Σ on its two faces along $\mathbf{N}$. $\bar{\mathbf{V}}$ is the material velocity of Σ, S is the entropy density, θ is the thermodynamical temperature, $\mathbf{P}$ is the material momentum, $\mathbf{b}$ is the Eshelby stress, $\mathbf{f}_\Sigma$ is the unknown driving force, and σ_Σ is the corresponding surface entropy at Σ. It was shown that for a coherent (no defects) homothermal (no temperature jump) front, we have

$$\mathbf{f}_\Sigma \cdot \bar{\mathbf{V}} = f_\Sigma \bar{V}_N = \theta_\Sigma \sigma_\Sigma \geq 0 \tag{9.83}$$

and

$$f_\Sigma = -\text{Hugo}_{PT}, \quad \text{Hugo}_{PT} := \mathbf{N} \cdot [\mathbf{b}_S] \cdot \mathbf{N} = [W - \langle \mathbf{N} \cdot \mathbf{T} \rangle \cdot \mathbf{F} \cdot \mathbf{N}], \tag{9.84}$$

where $\mathbf{b}_S$ is the quasi-static part of $\mathbf{b}$ (although the computation is made without neglecting inertia). If this inertia is really neglected, then we have the following reduction:

$$\text{Hugo}_{PT} = [W - tr(\langle \mathbf{T} \rangle \cdot \mathbf{F})]. \tag{9.85}$$

In this canonical formalism, the driving force $\mathbf{f}_\Sigma$ happens to be purely normal but it is constrained to satisfy, together with the propagation speed $\bar{V}_N$, the surface dissipation inequality indicated in the last term of (9.83). In other words, any relationship between these two quantities must be such that the inequality (9.83) be verified. This is the basis of the formulation of a *thermodynamically admissible criterion of progress* for Σ. Indeed, we should look for a relationship $\bar{V}_N = g(f_\Sigma)$ that satisfies the last term of (9.83). For illustrative purposes, we may consider the *cartoonesque* case where the phase transition process does not involve any characteristic time (just like rate-independent plasticity), in which case the dissipation (9.83) must be homogeneous of degree 1 only in $\bar{V}_N^\circ$; the threshold type of progress criterion corresponds to this. That is, $\bar{V}_N \in \partial\, I_f = N_C(f_\Sigma)$, where I_f is the indicator function of the closed segment $F=[-f_c, +f_c]$ – a convex set – and N_C is the "cone of outward normals" to this convex set, the symbol ∂ denoting the so-called subgradient (see Ref. [38], Appendix). If we "force" the system evolution to be such that there is effective progress of the front at $\mathbf{X} \in \Sigma$ while there is *no* dissipation, then we must necessarily enforce the following condition:

$$f_\Sigma = 0, \text{ that is, } \mathrm{Hugo}_{PT} \equiv [W - \langle \mathbf{N}\cdot\mathbf{T}\rangle \cdot \mathbf{F}\cdot\mathbf{N}] = 0. \tag{9.86}$$

On account of the fact that temperature (θ_Σ) is fixed and the thickness of the front is taken as zero, so that uniform states are reached immediately on both sides of Σ, Equation 9.86 is none other than the condition of "Maxwell" (9.78) in the one-dimensional pure shear case. Thus, a macroscopic approach dear to the engineer has allowed us to obtain, in general, a more realistic (in general, dissipative) progress of the front. The case of Section 9.4.2 then appears as a "*zoom*" – in the nondissipative case – on the situation described in the present section since the front acquires, through this zoom magnification (asymptotics), a definite, although small, thickness and a structure while rejecting the immediate vicinity of the zero-thickness front to infinities. The next approach allows one to introduce both *thickness* and *dissipation*.

9.4.4
Mesoscopic Applied Mathematics Approach: Structured Front

Here the front of phase transformation is looked upon as a mixed "*viscous dispersive*" structure at a mesoscale. We refer to this as the *applied mathematician approach*. This dialectical approach in which one applies macroscopic concepts at a smaller scale to obtain an improved phenomenological description is finally fruitful. Here, we refer to Truskinowsky [39] who considers a one-dimensional model (along the normal to the structured front, – cf. Figure 9.5) and envisages a *competition* between viscosity (i.e., a simple case of dissipation) and some *weak nonlocality* accounted for through a strain gradient theory. The critical nondimensional parameter that compares these two effects is defined by

$$\omega = \eta/\sqrt{\varepsilon}, \tag{9.87}$$

where η is the viscosity and $\varepsilon \approx L^2$ is the nonlocality parameter (size effect). Progressive wave solutions ($u = u(\xi = x - \bar{V}_N t)$) of the continuous system that relate two minimizers (uniform solutions at infinities that minimize $\bar{W}$) over a distance of the order of $\delta = \sqrt{\varepsilon}$ are discussed in terms of this parameter. The mathematical problem reduces to a *nonlinear eigenvalue problem* of which the specification of the points of the discrete spectrum constitutes the looked for *kinetic relation* $\bar{V}_N = g(f;\varepsilon)$, where $f = \bar{\sigma} - \bar{\sigma}(+\infty)$ plays the role of *driving force*. As a matter of fact, the speed of propagation $\bar{V}_N$ satisfies the *Rankine–Hugoniot equation* $\bar{V}_N^2 = [\bar{\sigma}]/[s]$, where strain gradients and viscosity play no role and the jumps are taken between asymptotic

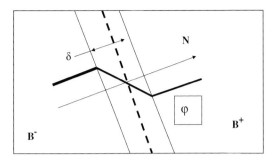

Figure 9.5 Phase transition front seen as a moving microstructured (viscous) transition zone, the thickness of which induces dispersion. The combination of viscosity and dispersion is responsible for the appearance of a moving structure subject to a driving "force" (the linear variation of a typical parameter – for example, gradient elasticity – is given as a simple illustration).

values at infinity (cf. Equation 9.79). The evolution obtained for the kinetic law is a strongly nonlinear function and evolves with the value of the parameter of ω.

9.4.5
Theoretical Physics Approach: Quasi-Particle and Transient Motion

The approach of Section 9.4.3 simply accepts the value of $\bar{V}_N$, whatever its evolution, as it is computed from the full field solution at each instant of time and each material point $\mathbf{X} \in \Sigma$. In contrast, the approaches of Sections 9.4.2 and 9.4.4 provide *progressive wave* solutions, that is, waves that are steady in the sense that the propagation speed, although a property of the solution (and *not only* of the material as in linear wave propagation), does not vary in time along the propagation path. This is a type of *inertial motion*. What about a *noninertial motion*? To look at such a case, one can view the problem as follows. The localized – but with nonzero thickness – dynamical solutions of Section 9.4.2 are looked upon as global entities (Figure 9.6) behaving like massive point particles in motion in the appropriate *point mechanics*, that is, as so-called *quasi-particles*. All perturbing effects such as dissipation, inhomogeneities, and

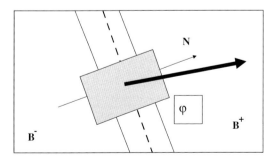

Figure 9.6 The phase transition front is seen in 1D propagation space as a more or less inertially moving "quasi-particle" (inertial motion in the absence of driving force).

so on will then be treated as *perturbing forces* acting on the inertial motion that thus becomes *noninertial*. To understand this viewpoint, it is sufficient to envisage the presence of a viscous (more generally, dissipative) contribution $\mathbf{f}_D$ on the right-hand side of the classical balance of linear momentum. This results in the presence of an additional *material force* $\mathbf{f}_D^{inh} = -\mathbf{f}_D \cdot \mathbf{F}$ on the right-hand side of the canonical momentum equation. The latter equation, after integration over the path of the wave, is used to treat this material force as a perturbation on the solution in the absence of $\mathbf{f}_D$. The essential problem then consists in identifying the *point mechanics* that is associated with a particular system of partial differential equations on account of some of its exact integrals. This point mechanics – that is, a coherent system of relations between mass, momentum, and energy of a point particle – can be completely new and a priori unforeseeable.

Remark. For the exploitation of thermodynamically based continuous automaton to treat numerically the progress of phase transition fronts in thermoelasticity, refer to Ref. [40].

9.4.6
Remark on a Mechanobiological Problem

Some physiological problems involving mechanics may look like problems dealing with phase transition fronts. This is the case of the growth of long bones under the influence of mechanical factors. Here, the main phenomenon is the growth at the so-called growth plate Σ that connects the metaphyseal bone and the epiphyseal bone [41]. This transition zone that may be called the "chondro-osseous junction" (from bone to cartilage) has a stationary motion that occurs with a competition between proliferation and hypertrophy of chondrocytes and ossification process. In spite of the complexity and multiplicity of processes involved in the activation of the different behaviors of the chondrocytes, the growth plate considered as a singular surface of vanishingly small thickness has a steady motion (during the lengthening of the bone that takes years) that is governed by a kinetic law such as

$$V_\Sigma = k f_\Sigma, \quad k > 0, \tag{9.88}$$

where (the μ's are chemical potentials)

$$b_N = \mathbf{N} \cdot [\mathbf{b}_S] \cdot \mathbf{N} = \mu_{bone} - \mu_{cart}, \quad \mathbf{b}_S = W \mathbf{1}_R - \mathbf{T} \cdot \mathbf{F}, \tag{9.89}$$

so that the local dissipation inequality $V_\Sigma f_\Sigma \geq 0$ and the surface balance $f_\Sigma + b_N = 0$ are satisfied. Bone and cartilage have different elastic potentials. $\mathbf{N}$ is the unit normal to the growth plate and both displacement and traction are continuous at Σ. The proportionality assumed in (9.88) is an oversimplification. The stability of the motion (9.88) can be studied. It is found that compression decreases the interface rate, while traction favors the lengthening of the bone (increase in V_Γ) as experimentally observed [41]. In analogy with the contents of previous sections, this problem can also be examined from different viewpoints (cf. Figure 9.7: (a) no-thickness transition discontinuity surface; (b) microstructured interface between the

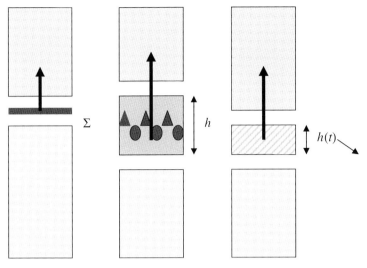

Figure 9.7 Progress of the growth plate in mechanobiology viewed in analogy with the propagation of phase transition fronts in inert matter.

two phases; (c) dissipative structure terminating in time with the almost complete disappearance of the growth plate).

9.5
Dynamic Materials

Those materials whose characteristic properties (e.g., mass density and elasticity) may be made to vary in time and possibly in space, or both, by an appropriate arrangement or control are called dynamic materials. Of course, materially inhomogeneous materials are known in various forms, polycrystals, composites of the stratified type, or so-called graded materials (with a more or less smooth gradient in their properties). We obviously ignore here inhomogeneous materials with stochastic properties. Materials inhomogeneous *in time* are not so frequent or are practically inexistent in natural conditions. We may conceive of some artificial means of causing these controlled changes in time, for example, by the application of an external (nonmechanical) field or through a phase transition. It is to be noted that this should be realized in a short or quasi-nil time lapse and over a sufficiently large material region if not over the whole specimen under consideration. Difficulties of realization cannot be overlooked. To be more specific, we consider the case of small strain elasticity, as introduced in Section 9.2.2. An especially interesting case is one where Equation 9.17 reduces to the following:

$$\frac{\partial}{\partial t}\left(\varrho_0(\mathbf{x})\frac{\partial \mathbf{u}}{\partial t}\right) - \operatorname{div} \boldsymbol{\sigma} = \mathbf{0}, \quad \boldsymbol{\sigma} = \frac{\partial \tilde{W}(\mathbf{e}; \tilde{t})}{\partial \mathbf{e}}, \quad \mathbf{e} := (\nabla \mathbf{u})_S. \quad (9.90)$$

For a one-dimensional motion with linear elasticity of Hooke's coefficient E, this yields the *linear* wave equation (cf. Equation 9.23):

$$u_{tt} - \hat{c}^2(x,t) u_{xx} = 0, \quad \hat{c}^2 = E(\bar{t})/\varrho_0(x). \tag{9.91}$$

It is readily checked that the conservation laws of energy and material momentum (9.24) and (9.25) are replaced by the following *inhomogeneous* equations:

$$\frac{\partial H}{\partial t} - \frac{\partial Q}{\partial x} = h := -\frac{\partial L}{\partial t}\bigg|_{\mathrm{expl}} \tag{9.92}$$

and

$$\frac{\partial P^f}{\partial t} - \frac{\partial b}{\partial x} = f := \frac{\partial L}{\partial x}\bigg|_{\mathrm{expl}}, \tag{9.93}$$

that is, neither energy nor canonical momentum are strictly conserved since we have the source terms

$$h = (W/E) E_{\bar{t}}, \quad f = (K/\varrho_0) \varrho_{0x}. \tag{9.94}$$

Equation 9.91 lends itself to some kind of space–time homogenization if some type of periodicity is assumed in Euclidean space–time. However, in view of quantities that should be conserved across space- and time-like discontinuities (see below) and the symmetry built in Equation 9.91, it might be preferable to rewrite the latter as two compatible first-order partial differential equations by introducing the auxiliary scalar field v so that

$$v_t = \sigma := E(\bar{t}) u_x, \quad v_x = p := \varrho_0(x) u_t. \tag{9.95}$$

Lurie [42] has studied the homogenization of system (9.95) and some asymptotic expansion of the u solution for long times compared to the period of the repeated motif (e.g., a checkerboard) in Euclidean 2D space–time.

Remark: In one space dimension, the original system (9.90) and (9.95) are symmetric in the interchange $x \to t$ and $E \to \varrho$. But this symmetry is an artifact of the one-dimensionality in space (and of the quadratic form of the energy of linear elasticity that compares exactly to the quadratic nature of the kinetic energy). If we want to avoid this pitfall, we need to consider two spatial dimensions at least in order to highlight the difference between scalar and vector-like quantities (see, for example, in the problem of reflection and refraction).

Now, the apparent simplicity of Equations 9.91 and 9.95 is misleading. Some dynamical solutions with space–time separate variables can be found to Equation 9.95 for smooth known variations of density and elasticity coefficients [43]. With appropriate variations, a concentration of energy can be observed although no nonlinearity is involved! But more to the point is the case of piecewise variations of these two quantities because this raises the question of what happens when a signal crosses a spatial interface Σ_x (at fixed x) or a time interface Σ_t (at fixed time). In the first case, we face a classical problem of wave propagation at the crossing of an interface separating two media of different acoustic properties (here due to change in density) with a convergence or divergence of waves by conservation of the momentum. In the second case, we face a problem akin to the one met in studying a Doppler-like effect with capture of energy from the outside and a resulting change in frequency. The phenomenon is in the same class as the Cerenkov effect, transition radiation, and so on [44]. The combination of the two in a true dynamic material may yield a concentration of energy although the system is fully linear. Numerical simulations support this behavior [43]. The analysis given by Lurie [42] with an asymptotic approach of a space–time homogenized checkerboard repeating the *same* motif in space–time (i.e., a spatiotemporal material composite with rectangular microstructure) shows that for a long time of propagation both energy and material momentum are conserved in the first asymptotic order (both Equations 9.92 and 9.93 in the limit with vanishing sources). To obtain the desired amplification effect, the successive cells must also show the appropriate oriented changes as indicated by the smooth variation case.

9.6
Further Extensions: Propagation in Metamaterials and Others

In previous sections, we have illustrated on special cases some consequences of nonlinearity, dispersion, and dissipation on the dynamics of materials, and more than often the competition or cooperation of these different properties. Save some exceptional situations (introduction of some viscosity), we did not consider cases where the mechanical constitutive equations are more complex than purely elastic ones. This was sufficient to pinpoint the essential dynamic properties that play a role in materials such as crystalline substances. The dynamical material systems with hysteresis are also of current interest (see Ref. [45] and many works by V. Gusev). The dynamical materials in the above-granted sense are somewhat artificial. We note that there are other cases that may be more rewarding if only in theory because of their originality and novelty. One such case is that of metamaterials that exhibit strange inertial properties. The so-called Maxwell–Rayleigh model introduced by Maugin [46] belongs to this class with a continuum in which are embedded small resonators that are linearly or nonlinearly coupled to a matrix. Other authors such as Milton and Willis [47] have exhibited some surprising negative inertia. Another case of present actuality is that of materials having really, or approaching the limit of, a fractal structure. Then this may require an appropriate generalization of the operators such as the Laplacian and the d'Alembertian using an adequate functional framework (in this regard, see Ref. [48]). As an illustrative example, we reproduce in Figure 9.8 the possible "dispersion relation" for a spatially 1D fractal elastic structure with fractal dimension $D = 1.5$.

To conclude, we have not developed all the mathematical techniques that are now at hand to study nonlinear dynamical effects because of the wealth of books offered on the market and dealing with that aspect. But materials at both engineering – meaning macroscopic scale – and now micro- and nanoscales offer a rather wide field of

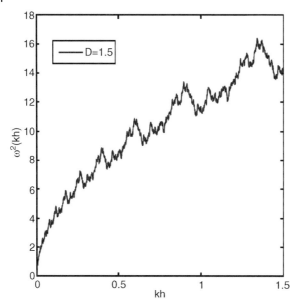

Figure 9.8 Typical dispersion relation for a linear fractal elastic chain of fractal dimension $D = 1.5$ (after Ref. [48]).

applications that are only superficially studied and are still in their infancy at the moment. The next two decades should witness a formidable development of the exploitation of all the emphasized dynamical properties, while enriching both the mathematical field and the numerical simulation techniques by their demand of efficient solutions.

References

1 Maugin, G.A. (2010) What do we understand by generalized continuum mechanics? in *Generalized Continuum Mechanics: One Hundred Years After the Cosserats* (eds G.A. Maugin and A.V. Metrikine), Springer, New York, pp. 3–13.

2 Maugin, G.A. (2006) On canonical equations of continuum thermomechanics. *Mech. Res. Commun.*, **33**, 705–710.

3 Maugin, G.A. (2000) Geometry of material space: its consequences in modern numerical means (FDM, FEM, Cellular Automata). *Tech. Mech.*, **20** (2), 95–104.

4 Maugin, G.A. (1993) *Material Inhomogeneities in Elasticity*, Chapman & Hall, London.

5 Brenig, W. (1955) Besitzen schallwellen einen impuls. *Z. Phys.*, **143**, 168–172.

6 Hayes, W.D. (1974) Introduction to wave propagation, in *Nonlinear Waves* (eds S. Liebowitz and A.R. Seebass), Cornell University Press, Ithaca, NY, pp. 1–43.

7 Maugin, G.A., Pouget, J., Drouot, R., and Collet, B. (1992) *Nonlinear Electromechanical Couplings*, John Wiley & Sons, Inc., New York.

8 Christov, C.I., Maugin, G.A., and Porubov, A.V. (2007) On Boussinesq's paradigm in nonlinear wave propagation. *C.R. Acad. Sci.*, **335** (9/10), 521–535.

9 Newell, A.C. (1985) *Solitons in Mathematics and Physics*, SIAM, Philadelphia, PA.

10. Kruskal, M.D. and Zabusky, N.J. (1966) Exact invariants for a class of nonlinear wave equations. *J. Math. Phys.*, **7**, 1265–1267.
11. Kruskal, M.D. (1974) The Korteweg–de Vries equation and related evolution equations, in *Nonlinear Wave Motion, Lectures in Applied Mathematics*, vol. 15 (ed. A.C. Newell), American Mathematical Society, Providence, RI, pp. 61–83.
12. Ablowitz, M.J. and Segur, H. (1981) *Solitons and the Inverse Scattering Transform*, SIAM, Philadelphia, PA.
13. Calogero, F. and Degasperis, A. (1982) *Spectral Transform and Solitons: Tools to Solve and Investigate Nonlinear Evolution Equations*, vol. I, North-Holland, Amsterdam.
14. Fokas, A.S. (1979) Generalized symmetries and constant of motion of evolution equations. *Lett. Math. Phys.*, **3**, 467–473.
15. Maugin, G.A. and Christov, C.I. (2002) Nonlinear duality between elastic waves and quasi-particles, in *Selected Topics in Nonlinear Wave Mechanics* (eds C.I. Christov and A. Guran), Birkhäuser, Boston, MA, pp. 101–145.
16. Rebi, C. and Soliani, G. (eds) (1984) *Solitons and Particles*, World Scientific, Singapore.
17. Holland, P.R. (1983) *The Quantum Theory of Motion*, Cambridge University Press, UK.
18. Sanz-Serna, J.M. and Calvo, M.P. (1994) *Numerical Hamiltonian Problems*, Chapman & Hall, London.
19. Pouget, J. (1988) Nonlinear dynamics of lattice models for elastic continua, in *NATO Summer School on Physical Properties and Thermodynamical Behavior of Minerals (Oxford, 1988)* (ed. K. Saljé), Reidel, Dordrecht, pp. 359–402.
20. Christov, C.I. and Maugin, G.A. (1993) Long-time evolution of acoustic signals in nonlinear crystals, in *Advances in Nonlinear Acoustics* (ed. H. Hobaek), World Scientific, Singapore, pp. 457–462.
21. Bogdan, M.M., Kosevich, A.M., and Maugin, G.A. (1999) Formation of soliton complexes in dispersive systems. *J. Phys. Condens. Matter*, **2** (2), 255–265.
22. Christov, C.I. and Maugin, G.A. (1995) An implicit difference scheme for the long time evolution of localized solutions of a generalized Boussinesq system. *J. Comput. Phys.*, **116**, 39–51.
23. Maugin, G.A. (1999) *Nonlinear Waves in Elastic Crystals*, Oxford Mathematical Monographs, Oxford University Press, UK.
24. Maugin, G.A. and Cadet, S. (1991) Existence of solitary waves in martensitic alloys. *Int. J. Eng. Sci.*, **29**, 243–255.
25. Christiansen, P.L. and Olsen, O.H. (1982) Propagation of fluxons on Josephson lines with impurities. *Wave Motion*, **4**, 163–172.
26. Wesolowski, Z. (1983) Dynamics of a bar of asymmetric cross section. *J. Eng. Math.*, **17**, 315–322.
27. Frenkel, Y.I. and Kontorova, TA. (1938) On the theory of plastic deformation and twinning. *Phys. Z. Sowjetunion*, **123**, 1–15.
28. Maugin, G.A. and Miled, A. (1986) Solitary waves in elastic ferromagnets. *Phys. Rev.*, **B33**, 4830–4842.
29. Pouget, J. and Maugin, G.A. (1984) Solitons and electroacoustic interactions in ferroelectric crystals. I. Single solitons and domain walls. *Phys. Rev. B*, **30**, 5306–5325.
30. Kivshar, Y.S. and Malomed, B.A. (1989) Dynamics of solitons in nearly integrable systems. *Rev. Mod. Phys.*, **61**, 763–915.
31. Pouget, J. and Maugin, G.A. (1985) Solitons and electroacoustic interactions in ferroelectric crystals. II. Interactions of solitons and radiations. *Phys. Rev.*, **B31**, 4633–4649.
32. Pouget, J. and Maugin, G.A. (1989) Nonlinear dynamic of oriented solids. I, II. *J. Elast.*, **22**, 135–156, 157–183.
33. Maugin, G.A. and Miled, A. (1986) Solitary waves in micropolar elastic crystals. *Int. J. Eng. Sci.*, **24**, 1477–1499.
34. Falk, F. (1983) Ginzburg–Landau theory of static domain walls in shape-memory alloys. *Z. Phys. C: Condens. Matter*, **51**, 177–185.
35. Maugin, G.A. and Trimarco, C. (1995) The dynamics of configurational forces at phase-transition fronts. *Meccanica*, **30**, 605–619.
36. Maugin, G.A. (1997) Thermomechanics of inhomogeneous–heterogeneous systems: application to the irreversible

progress of two- and three-dimensional defects. *ARI*, **50**, 41–56.

37 Maugin, G.A. (1998) On shock waves and phase-transition fronts in continua. *ARI*, **50**, 141–150.

38 Maugin, G.A. (1992) *Thermomechanics of Plasticity and Fracture*, Cambridge University Press, Cambridge, MA.

39 Truskinowsky, L.M. (1994) About the normal growth approximation in the dynamical theory of phase transitions. *Continuum Mech. Thermodyn.*, **6**, 185–208.

40 Berezovski, A., Engelbrecht, J., and Maugin, G.A. (2008) *Numerical Simulation of Waves and Fronts in Inhomogeneous Solids*, World Scientific, Singapore.

41 Sharipova, L., Maugin, G.A., and Freidin, A.B. (2008) Modelling the influence of mechanical factors on the growth plate. International Conference on Nonlinear Waves, Hong Kong, June 2008.

42 Lurie, K.A. (2007) *An Introduction to the Mathematical Theory of Dynamic Materials*, Springer, New York.

43 Rousseau, M., Maugin, G.A., and Berezovski, M. (2010) Elements of study on dynamic materials. *Arch. Appl. Mech.*, AAM-10-0040-R2 (D-RW-01).

44 Ginzburg, V.L. and Tsytovich, V.N. (1979) Several problems of the theory of transition radiation and transition scattering. *Phys. Rep.*, **49**, 1–89.

45 Ostrosvky, L.A. and Johnson, P.A. (2001) Dynamic nonlinear elasticity in geomaterials. *Riv. Nuovo Cimento*, **24**/7, 1–46.

46 Maugin, G.A. (1995) On some generalizations of Boussinesq and KdV systems. *Proc. Acad. Sci. Estonia*, **A44**, 40–55 (Special Issue: KdV equation).

47 Milton, G.W. and Willis, J.R. (2007) On modifications of Newton's second law and linear continuum elastodynamics. *Proc. R. Soc. Lond. A*, **463**, 855–880.

48 Michelitsch, T.M., Maugin, G.A., Nowakowski, A.F., Nicolleau, F.C.G.A., and Deroga, S. (2009) Dispersion relations and wave operators in self-similar quasi-continuous linear chains. *Phys. Rev. E*, **80**, 011135-1–011135-8.

10
Synchronization on the Circle
Alain Sarlette and Rodolphe Sepulchre

10.1
Introduction

Investigations on complex dynamical systems have been centered on two major topics in the last decades. The study of chaotic phenomena has been continuously pursued since Lorentz's famous equations [1], leading, for example, to applications in signal processing like in Ref. [2] and to laser physics in this book (Chapters 3 and 4). In parallel, spontaneous ordering – "synchronization" – phenomena have been observed in a wide variety of natural systems [3], including animal groups, coupled biological cells and physical systems (see, for example, "phase transitions" toward order [4]). These two apparently distinct topics have met in the study of chaotic oscillator synchronization (e.g., Ref. [5] and Chapter 4).

The occurrence of synchronization in natural systems is diversified, ranging, for example, from agreement in multiagent systems (animal groups) to the generation of rhythmic behavior at macroscopic scale through oscillator synchronization (pacemaker cells). Like chaos, synchronization was first studied in the physics literature and has only recently been picked up for engineering applications. Many recent engineering applications consider swarms of agents that combine their efforts in a coordinated ("synchronized") way to achieve a common task, for example, distributed exploration [6, 7] or interferometry [8, 9]. From a modeling and analysis viewpoint, collective phenomena are studied by both the theoretically oriented communities of dynamical systems and statistical physics and the more practically oriented experimental physics and biology communities (see, for example, [3, 4, 10, 11]). Celebrated examples of simplified models for the study of coordination phenomena include Kuramoto and Vicsek models.

Kuramoto model is proposed in Ref. [11] to describe the continuous time evolution of *phase variables* in a population of weakly coupled oscillators. Each agent k is considered as a periodic oscillator of natural frequency ω_k, whose phase – that is, position on its cycle – at time t is $\theta_k(t) \in S^1$. In addition to its natural evolution, each

---

The Complexity of Dynamical Systems. Edited by J. Dubbeldam, K. Green, and D. Lenstra
Copyright © 2011 WILEY-VCH Verlag GmbH & Co. KGaA, Weinheim
ISBN: 978-3-527-40931-0

agent is coupled to all the others:

$$\frac{d\theta_k}{dt} = \omega_k + \sum_{k=1}^{N} \sin(\theta_j - \theta_k), \quad k = 1, 2, \ldots, N. \quad (10.1)$$

Vicsek model is proposed in Ref. [4] to describe the discrete-time evolution of interacting particles that move with unit velocity in the plane. In the absence of noise, the model writes

$$r_k(t+1) = r_k(t) + e^{i\theta_k(t)}, \quad (10.2)$$

$$\theta_k(t+1) = \arg\left(\sum_{j \in n_k(t)} e^{i\theta_j(t)} + e^{i\theta_k(t)}\right), \quad k = 1, 2, \ldots, N, \quad (10.3)$$

where θ_k denotes the heading angle of particle k and r_k its position in the (complex) plane. The set $n_k(t)$ is defined to contain all agents j for which $\|r_k(t) - r_j(t)\| \leq R$ for some fixed $R > 0$, such that agent k is influenced only by "close enough" fellows.[1]

In the engineering community, synchronization has focused on several basic problems. One of them pertains to a set of agents to reach agreement on some "quantity" – for example, a position, velocity, and so on – while exchanging information along a limited set of communication links. This so-called *consensus problem* has been extensively studied for quantities on the *vector space* $\mathbb{R}^n$ (see, for example, Refs [12–18] for a review); related results are briefly explained in the next section. However, it appears that many interesting engineering applications and phenomenological models, for example, as presented above, involve nonlinear *manifolds*; this includes distributed exploration of a planet (sphere S^2), rigid body orientations (special orthogonal groups SO(2) or SO(3)), and collective motion on limit cycles (abstracted as a circle S^1).

Vicsek and Kuramoto models illustrate two general types of problems for synchronization on manifolds. In Vicsek model, stabilizing parallel motion requires equal orientations of the agents, which comes down to reaching *agreement on states* θ_k on the circle. In Kuramoto model, moving with fixed relative phases corresponds to reaching *agreement on a motion* on the circle. On multidimensional manifolds, coordinated motion becomes more complex than on S^1 where it simply comes down to reaching agreement on a frequency in $\mathbb{R}$; this is the subject of Ref. [19]. This chapter focuses on state agreement, more commonly called *synchronization*.

Definition 10.1

A swarm of N agents with states $x_k(t) \in \mathcal{M}$ evolving on a manifold $\mathcal{M}$, $k = 1, 2, \ldots, N$, is said to asymptotically synchronize or reach synchronization if

---

1) Notation $\|z\|$ denotes the complex norm of $z \in \mathbb{C}$, such that $z = \|z\|e^{i\arg(z)}$. This chapter uses the convention that $\arg(0)$ can take any value on S^1. It canonically identifies $\mathbb{C} \cong \mathbb{R}^2$, yielding $\|z\| = \sqrt{z^T z}$ for $z \in \mathbb{R}^2$.

$x_1 = x_2 = \cdots = x_N$ asymptotically.

The asymptotic value of x_k is called the consensus value.

The circle is probably the simplest nonlinear manifold to highlight the specificities of synchronization on highly symmetric compact nonlinear spaces. Therefore, this chapter focuses on the circle and mentions extensions to manifolds at the end. An important aspect is to maintain the *symmetry* of the synchronization problem: the behavior of the swarm must be invariant with respect to a common translation of all the agents (i.e., on S^1, under the transformation $\theta_k \to \theta_k + a \ \forall k \in \mathcal{V}$); therefore, extensions consider manifolds on which "all points are equivalent" – formally, *compact homogeneous manifolds* like S^n, $SO(n)$, or the Grassmann manifolds. In order to concentrate on fundamental issues of synchronization and geometry, system dynamics are simplified to first-order integrators.

The chapter is organized as follows. Section 10.2 briefly reviews a fundamental algorithm for synchronization on vector spaces, also known as a "consensus algorithm," in continuous and discrete time. Section 10.3 extends this algorithm to the circle and highlights its link with Kuramoto [11] and Vicsek [4] models. Section 10.4 reviews the positive synchronization properties of these algorithms. Section 10.5 illustrates the fact that, in contrast to vector spaces, convergence to nonsynchronized behavior is possible on the circle. Section 10.6 presents three recently proposed algorithms for (almost) global synchronization on the circle: a *modified coupling function*, a *gossip algorithm*, where agents randomly select or discard information from their neighbors, and an algorithm based on auxiliary variables. Section 10.7 briefly mentions extensions of the framework to compact homogeneous manifolds. General notations and background information about graphs can be found in Appendix 10.A. Proofs are summarized to their main ideas; complete versions can be found in corresponding references and in Ref. [20].

10.2
Consensus Algorithms on Vector Spaces

The study of synchronization on vector spaces is a widely covered subject in the systems and control literature of the decade gone by. In this section, we present a summary of basic results achieved in this framework by several authors, including Refs [12–14, 16–18]; see Ref. [15] for a review and Ref. [14] for some examples of applications.

Consider a *swarm* of N agents with states $x_k \in \mathbb{R}^n$, $k \in \mathcal{V} = \{1, 2, \ldots, N\}$, evolving under continuous-time dynamics

$$\frac{d}{dt} x_k(t) = u_k(t), \quad k = 1, 2 \ldots, N \tag{10.4}$$

or discrete-time dynamics

$$x_k(t+1) = x_k(t) + u_k(t), \quad k = 1, 2 \ldots, N, \tag{10.5}$$

where u_k is a coupling term. The goal is to design u_k such that the agents asymptotically synchronize in the sense of Definition 10.1, with the following restrictions:

1) **Communication constraint.** u_k may depend only on information concerning agent k and the agents $j \rightsquigarrow k$ sending information to k according to some imposed communication graph $\mathbb{G}$ (see Appendix 10.A).
2) **Configuration space symmetry.** The behavior of the coupled swarm must be invariant with respect to uniform translation of all the agents: defining $y_k(0) = x_k(0) + a \ \forall k \in \mathcal{V}$ for any $a \in \mathbb{R}^n$, it must hold $y_k(t) = x_k(t) + a \ \forall k \in \mathcal{V}$ and $\forall t \geq 0$. Therefore, u_k may depend only on the *relative* positions of the agents, that is, on $(x_j - x_k)$ for $j \rightsquigarrow k$.
3) **Agent equivalence symmetry.** All the agents in the swarm must be treated equivalently. This implies that (i) the form of u_k must be the same $\forall k \in \mathcal{V}$ and (ii) all $j \rightsquigarrow k$ must be treated equivalently in u_k.

This problem is traditionally called the *consensus problem on a vector space*. On manifolds, the term "synchronization problem" is preferred because the term "consensus" can be given a particular meaning different from synchronization (see Ref. [21]). On vector spaces, consensus as defined in Ref. [21] is equivalent to synchronization, so both terms can be used interchangeably.

The consensus problem on vector spaces is solved by the linear coupling

$$u_k(t) = \alpha \sum_{j=1}^{N} a_{jk}(t)(x_j(t) - x_k(t)), \quad k = 1, 2 \ldots, N, \tag{10.6}$$

where a_{jk} is the weight of link $j \rightsquigarrow k$ and α is a positive gain. The intuition behind (10.6) is that each agent moves toward its neighbors, in agreement with the traditional meaning given to a consensus process. In continuous time, the closed loop system (10.4) and (10.6) implies that agent k is moving toward the position in $\mathbb{R}^n$ corresponding to the (positively weighted) arithmetic mean of its neighbors, $\left(1/d_k^{(i)}\right)\sum_{j \rightsquigarrow k} a_{jk} x_j$, where in-degree $d_k^{(i)} = \sum_{j \rightsquigarrow k} a_{jk}$. In discrete time, α must satisfy $\alpha \, d_k^{(i)}(t) \leq b$ for some constant $b < 1$, $\forall k \in \mathcal{V}$. Then, (10.5) and (10.6) mean that the future position of agent k is at the (positively weighted) arithmetic mean $\left(1/(\beta_k + d_k^{(i)})\right)\left(\sum_{j \rightsquigarrow k} a_{jk} x_j + \beta_k x_k\right)$ of its neighbors $j \rightsquigarrow k$ and itself, with nonvanishing weight β_k.

Clearly, (10.6) satisfies the three constraints mentioned above.

The convergence properties of the linear consensus algorithm on a vector space are well characterized. An extension of the following basic result in the presence of time delays can be found in Ref. [16]; this chapter does not consider time delays.

Proposition 10.1

(Adapted from Refs [13, 14, 16, 18]) *Consider a set of N agents evolving on $\mathbb{R}^n$ according to (continuous time) (10.4) and (10.6) with $\alpha > 0$ or according to (discrete time) (10.5)* and (10.6) with $\alpha d_k^{(i)}(t) \in [0, b]\ \forall t \geq 0$ and $\forall k \in \mathcal{V}$, for some constant $b \in (0, 1)$. Then, the agents globally and exponentially converge to synchronization at some constant value $\bar{x} \in \mathbb{R}^n$ if the communication among agents is characterized by a (piecewise continuous) δ-digraph that is uniformly connected.

If, in addition, $\mathbb{G}$ is balanced for all times, then the consensus value is the arithmetic mean of the initial values: $\bar{x} = (1/N)\sum_{k=1}^{N} x_k(0)$.

Proof idea

For the first part, see Refs [13, 14], or equivalently Refs [12, 18]. For the second part, it is easy to see that for a balanced graph, $(1/N)\sum_{k=1}^{N} x_k(t)$ is conserved over time. The conclusion is then obtained by comparing its value for $t = 0$ and for t going to $+\infty$.

The proof of Proposition 10.1 essentially relies on the *convexity* of the update law: the position of each agent k for $t > \tau$ always lies in the convex hull of the $x_j(\tau)$, $j = 1, 2, \ldots, N$. The permanent contraction of this convex hull, at some nonzero minimal rate because weights are nonvanishing, allows to conclude that the agents end up at a consensus value. An obvious negative counterpart of Proposition 10.1 for nonvarying $\mathbb{G}$ is that synchronization cannot be reached if $\mathbb{G}$ is not root connected.

If interconnections are not only balanced but also undirected and fixed, then the linear consensus algorithm is a gradient descent algorithm for the disagreement cost function

$$V_{\text{vect}}(x) = \frac{1}{2}\sum_{k=1}^{N}\sum_{j=1}^{N} a_{jk}\|x_j - x_k\|^2 = \|(B \otimes I_n)x\|^2 = x^T(L \otimes I_n)x, \quad (10.7)$$

where $\|z\|$ denotes the Euclidean norm $\sqrt{z^T z}$ of $z \in \mathbb{R}^m$, B and L are the incidence and Laplacian matrices of $\mathbb{G}$, respectively (see Appendix 10.A), $x \in \mathbb{R}^{Nn}$ denotes the vector whose elements $(k-1)n + 1$ to kn contain x_k, and $\otimes I_n$ is the Kronecker product by the $n \times n$ identity matrix.

10.3
Consensus Algorithms on the Circle

Consider a swarm of N agents with states on the circle S^1. The global topology of the circle is fundamentally different from vector spaces because if θ_k denotes an angular position on the circle, then $\theta_k + 2\pi = \theta_k$, that is, translations on the circle are defined modulo 2π because they correspond to rotations. This difference in topology, imposing a nonconvex configuration space, fundamentally modifies the synchronization problem.

The synchronization problem on S^1 is considered under the same agent dynamics as on $\mathbb{R}^n$, that is, (10.4) or (10.5) with x_k replaced by θ_k. However, for the design of u_k, the different topology induces different implications of the configuration space symmetry. The behavior of the swarm must (i) be invariant with respect to a uniform translation of all θ_k and (ii) be invariant with respect to the translation of any single

θ_k by a multiple of 2π – that is, if $\phi_k(0) = \theta_k(0) + 2a\pi$ for some $k \in \mathcal{V}$ and $a \in \mathbb{Z}$, and $\phi_j(0) = \theta_j(0) \ \forall j \neq k$, then it must hold $\phi_k(t) = \theta_k(t) + 2a\pi \ \forall t \geq 0$ and $\phi_j(t) = \theta_j(t) \ \forall j \neq k$ and $\forall t \geq 0$. This implies that u_k may depend only on 2π-periodic functions of the relative positions $(\theta_j - \theta_k)$ of the agents $j \rightsquigarrow k$. The simple linear algorithm (10.6) does not satisfy the periodicity required for configuration space symmetry and therefore cannot be used on the circle. It can, however, be used to derive algorithms for synchronization on S^1 that are similar to (10.6) when all agents are within a small arc of the circle. The discrete-time and continuous-time cases are treated consecutively. Because of the symmetry with respect to uniform translations on S^1, the swarm's behavior is entirely characterized by examining the evolution of *relative* positions.

Definition 10.2

A configuration is a particular set of relative positions of the agents. Thus, a configuration is equivalent to a point $(\bar\theta_1, \bar\theta_2, \ldots, \bar\theta_N) \in S^1 \times S^1 \times \cdots \times S^1$ and all the points obtained by its uniform rotations $(\bar\theta_1 + a, \bar\theta_2 + a, \ldots, \bar\theta_N + a)$ for $a \in S^1$.

10.3.1
Discrete Time

Synchronization of $\theta_k \in S^1$, $k = 1, 2, \ldots, N$, can be seen as synchronization of $x_k \in \mathbb{R}^2$ under the constraint $\|x_k\| = 1$. If the x_k were not restricted to $\|x_k\| = 1$, algorithm (10.5) and (10.6) would impose $x_k(t+1) = (1/(\beta_k + d_k^{(i)})) \left(\sum_{j=1}^N a_{jk} x_j(t) + \beta_k x_k(t) \right)$, $k = 1, 2, \ldots, N$, with some nonvanishing $\beta_k(t) > 0$. With this update law, $x_k(t+1)$ does generally not satisfy $\|x_k(t+1)\| = 1$. To obtain $\|x_k(t+1)\| = 1$, the result of algorithm (10.5) and (10.6) is projected onto the unit circle. Identifying $\mathbb{R}^2 \cong \mathbb{C}$, such that a position on the circle is characterized by $e^{i\theta_k}$, leads to the discrete-time synchronization algorithm

$$\theta_k(t+1) = \arg\left(\sum_{j=1}^N a_{jk} e^{i\theta_j(t)} + \beta e^{i\theta_k(t)} \right), \quad k = 1, 2, \ldots, N, \quad (10.8)$$

for some constant $\beta > 0$. The update of one agent according to (10.8) is illustrated in Figure 10.1. It is clear from the figure that (10.8) respects the geometric invariance of S^1. This is confirmed by rewriting (10.8) as

$$\theta_k(t+1) = \theta_k(t) + u_k = \theta_k(t) + \arg\left(\sum_{j=1}^N a_{jk} e^{i(\theta_j(t) - \theta_k(t))} + \beta \right), \quad k = 1, 2, \ldots, N, \quad (10.9)$$

where u_k indeed involves only 2π-periodic functions of relative positions of connected agents $j \rightsquigarrow k$.

For fixed undirected $\mathbb{G}$, the point $\theta_k(t+1)$ obtained from (10.8) is the projection on the unit circle of a point obtained by gradient descent for V_{vect} in the complex plane.

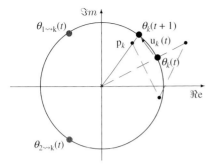

Figure 10.1 Illustration of update law (10.8) for one agent k with $\beta = 1.5$ and $a_{1k} = a_{2k} = 1$.

10.3.1.1 Vicsek Model

Heading update law (10.3) actually corresponds to (10.8) with $\beta = 1$ and $a_{jk} \in \{0, 1\}$. The positions define the interconnection graph $\mathbb{G}(t)$ by imposing $a_{jk} = 1$ if and only if $\|r_k(t) - r_j(t)\| \leq R$; therefore, $\mathbb{G}$ is called a *proximity graph*. The study of proximity graphs, or other state-dependent graphs, is beyond the scope of this chapter.

10.3.2
Continuous Time

Taking the continuous-time limit of (10.8) amounts to letting β grow indefinitely. In this limit case, $x_k(t+1) \in \mathbb{C}$ is defined with an infinitesimal gradient step for V_{vect}, and projected onto S^1 to yield $\theta_k(t+1)$. This is strictly equivalent to projecting the gradient of V_{vect} onto the tangent to the circle at $\theta_k(t)$ and taking a corresponding infinitesimal descent step along the circle. Thus, by viewing V_{vect} as a function of θ, renamed for clarity

$$V_{\text{circ}}(\theta) = \frac{1}{2}\sum_{k=1}^{N}\sum_{j=1}^{N} a_{jk}\|e^{i\theta_j} - e^{i\theta_k}\|^2 = \frac{1}{2}\sum_{k=1}^{N}\sum_{j=1}^{N} a_{jk}\left(2\sin\left(\frac{\theta_j - \theta_k}{2}\right)\right)^2, \quad (10.10)$$

the corresponding gradient descent algorithm along the circle, $d\theta_k/dt = -\alpha(\partial V_{\text{circ}}/\partial \theta_k) \forall k \in \mathcal{V}$ with $\alpha > 0$, is the continuous-time limit of (10.8). Computing the gradient of (10.10) yields the following continuous-time algorithm for synchronization on the circle, with constant $\alpha > 0$:

$$\frac{d}{dt}\theta_k = \alpha \sum_{j=1}^{N}(a_{jk}(t) + a_{kj}(t))\sin(\theta_j(t) - \theta_k(t)), \quad k = 1, 2, \ldots, N. \quad (10.11)$$

This algorithm can be implemented only for undirected $\mathbb{G}$. An extension to directed graphs is

$$\frac{d}{dt}\theta_k = 2\alpha \sum_{j=1}^{N} a_{jk}(t)\sin(\theta_j(t) - \theta_k(t)), \quad k = 1, 2, \ldots, N. \quad (10.12)$$

This is the actual algorithm considered in the following. It satisfies the geometric invariance of S^1 since the right-hand side is a 2π-periodic function of relative positions. With $x_k = e^{i\theta_k}$, (10.12) is equivalent to

$$\frac{d}{dt} x_k = 2\alpha \, \text{proj}_{x_k} \left(\sum_{j=1}^{N} a_{jk}(x_j - x_k) \right), \tag{10.13}$$

where $\text{proj}_{x_k}(r_k) = r_k - x_k x_k^T r_k$ denotes the orthogonal projection of $r_k \in \mathbb{C} \cong \mathbb{R}^2$ onto the direction tangent to the unit circle at $x_k = e^{i\theta_k}$. The geometric interpretation is that (10.13) defines a consensus update similar to (10.4) and (10.6) but constrained to the subset of $\mathbb{R}^2$ where $\|x_k\| = 1$. Algorithm (10.12) was proposed in Ref. [22] in a control framework and is directly linked to Kuramoto model.

10.3.2.1 Kuramoto Model

Comparing with (10.1), algorithm (10.12) in fact corresponds to Kuramoto model with equal natural frequencies $\omega_1 = \omega_2 = \cdots = \omega_N$, but general interconnections. This highlights a link between the sine-model of Kuramoto and the "averaging" update law for headings in Vicsek model. For the complete graph, $V_{\text{circ}} = 1/2 \sum_{k=1}^{N} \sum_{j=1}^{N} \|e^{i\theta_j} - e^{i\theta_k}\|^2 = N^2 - \|\sum_{k=1}^{N} e^{i\theta_k}\|^2$. The quantity $\|\sum_{k=1}^{N} e^{i\theta_k}\|^2$, known as the "complex order parameter" in the context of Kuramoto model, has been used for decades as a measure of the synchrony of phase variables in the literature on coupled oscillators.

The main point in studies of Kuramoto model is the coordination of agents having different ω_k. The important issue of robustly coordinating agents despite their different natural tendencies is not the subject of this chapter.

10.4
Convergence Properties

Section 10.3 proposes algorithms (10.8) and (10.12) as natural extensions of synchronization algorithms for the circle. However, the circle is not a convex configuration space. As a consequence, the convergence properties of (10.8) and (10.12) do not match those of (10.4)–(10.6) on vector spaces. This section focuses on positive convergence results, while Section 10.5 focuses on situations in which asymptotic synchronization is not achieved.

10.4.1
Local Synchronization Like for Vector Spaces

When all agents are within a small subset of S^1, (10.12) becomes similar to (10.4) and (10.6) because $\sin(\theta_j - \theta_k) \simeq (\theta_j - \theta_k)$ for small $(\theta_j - \theta_k)$. A similar observation can be made for the discrete-time algorithms. It is thus not surprising that [14, 23] are able to show that asymptotic synchronization on the circle is locally achieved under the same conditions as in vector spaces.

Proposition 10.2

(Adapted from Ref. [14]) *Consider a set of N agents evolving on S^1 according to (continuous time) (10.12) with $\alpha > 0$ or according to (discrete time) (10.8) with $\beta > 0$. If the communication among agents is characterized by a (piecewise continuous) δ-digraph $\mathbb{G}$, which is uniformly connected, and all agents are initially located within an open semicircle, then they exponentially converge to synchronization at some constant value $\bar{\theta} \in S^1$.*

Proof idea

Assume without loss of generality that $\theta_k \in [-b, b] \subset (-\pi/2, \pi/2)$ initially. Then, $\sin(\theta_k(t) - \theta_j(t)) = c_{jk}(t)(\theta_j(t) - \theta_k(t))$, where $c_{jk}(t) \geq \sin(2b)/2b > 0$ depends on $(\theta_j(t) - \theta_k(t))$. Thus, (10.12) is equivalent to

$$\frac{d}{dt}\theta_k = \alpha \sum_{j=1}^{N} a_{jk}(t) c_{jk}(t) (\theta_j(t) - \theta_k(t)) \tag{10.14}$$

for some time-varying $c_{jk} \geq \sin(2b)/2b > 0$. This can be viewed as a linear synchronization algorithm for $\theta \in \mathbb{R}$ with a δ_2-digraph $\mathbb{G}_2$ of weights $(a_{jk} c_{jk})$. A similar idea can be used in discrete time.

When the agents are distributed over more than a semicircle, the proof of Proposition 10.2 no longer holds because the c_{jk} can be negative. This is, in fact, the consequence of *loss of convexity*, implying that the strong vector space arguments of Refs [13, 14] are no longer applicable.[2] In algorithms (10.8) and (10.12), agents move on the shortest path toward their neighbors. Therefore, if the agents are initially located within a semicircle, they remain within this set for all future times, while agents distributed over more than a semicircle may, *a priori*, leave any open arc $s \in S^1$ containing them all. A more global analysis requires stronger assumptions.

10.4.2
Some Graphs Ensure (Almost) Global Synchronization

For general graphs, synchronization is only locally asymptotically stable. For some graphs, however, synchronization is (almost) globally asymptotically stable.

Proposition 10.3

(Adapted from Refs [24, 25]) *Consider a set of N agents evolving on S^1 by applying (10.12) with $\alpha > 0$ or (10.8) with $\beta > 0$ not too small (see Propositions 10.5 and 10.6 in the following section). If communication graph $\mathbb{G}$ is a fixed directed root-connected tree, or an undirected tree, or a complete graph, or any vertex interconnection of trees and*

---

[2] An open subset $s \subset S^1$ is *convex* if it contains all shortest paths between any two points of s.

complete graphs,[3] then the agents asymptotically converge to synchronization, for almost all initial conditions.

Proof idea

For the directed rooted tree, each agent is attracted toward its parent and, except for unstable situations where two connected agents are exactly at opposite positions on the circle, they synchronize at the initial position of the root. The particular undirected graphs have the property that synchronization is the only local minimum of V_{circ}. This is rather obvious for the tree, from the fact that for any pair of connected agents, variations of V_{circ} can be built involving only the distance between that particular pair of agents. The property is proved for the complete graph in Ref. [25]. Finally, Ref. [24] shows that the property holds for vertex interconnections of graphs for which it holds individually. Then, Propositions 10.4–10.6 of the following section ensure that synchronization is the only stable limit set.

10.5
Obstacles to Global Synchronization

Section 10.4 identifies situations where (10.8) and (10.12) converge to synchronization. The present section examines what can happen when this is not the case.

10.5.1
Convergence to Local Equilibria for Fixed Undirected $\mathbb{G}$

The fact that for fixed undirected $\mathbb{G}$ (10.12) is a gradient descent for V_{circ} has strong implications for the convergence analysis.

Proposition 10.4

Consider a set of N agents evolving on S^1 according to (10.12) with $\alpha > 0$, with communication graph $\mathbb{G}$ fixed and undirected. Then, the agents always converge to a set of equilibria corresponding to the critical points[4] of V_{circ} defined in (10.10). The only asymptotically stable equilibria are the local minima of V_{circ}.

Proof idea

Properties of gradient algorithms.

---

3) A vertex interconnection of two graphs $\mathbb{G}_1(\mathcal{V}_1, \mathcal{E}_1)$ and $\mathbb{G}_2(\mathcal{V}_2, \mathcal{E}_2)$ is a graph $\mathbb{G}$ whose vertices can be partitioned into a singleton $\{k\}$ and two sets $\mathcal{V}_a, \mathcal{V}_b$ and whose edge set can be partitioned into two sets $\mathcal{E}_a, \mathcal{E}_b$, such that $\mathcal{V}_a \cup \{k\} = \mathcal{V}_1$, $\mathcal{E}_a = \mathcal{E}_1$, $\mathcal{V}_b \cup \{k\} = \mathcal{V}_2$, and $\mathcal{E}_b = \mathcal{E}_2$.
4) A *critical point* of a differentiable function $f : \mathcal{X} \to \mathbb{R}$ is a point of $\mathcal{X}$, where the gradient of f is identically zero.

For the discrete-time algorithm, in addition to a result similar to Proposition 10.4 but with a bound on β, it can be shown that under *locally asynchronous update*, convergence holds for arbitrary β, that is, without requiring any minimal inertia. Agents are said to update *synchronously* if, between instants t and $t+1$, all agents $k \in \mathcal{V}$ apply (10.8). In contrast, agents are said to update *locally asynchronously* if, between instants t and $t+1$, only a subset of agents $\sigma \subset \mathcal{V}$ applies (10.8) and the others remain at their position, and set σ contains no agents connected to each other in $\mathbb{G}$.

Proposition 10.5

(Adapted from Ref. [26]) *Consider a set of N agents evolving on S^1 by applying (10.8) locally asynchronously with update subset sequence $\sigma(t)$, for $\beta > 0$ and fixed undirected $\mathbb{G}$. Assume that there exist a finite time span T and a partition of the discrete-time space $[t_0, t_1), [t_1, t_2), \ldots$, with $(t_{n+1} - t_n) < T \; \forall n \in \mathbb{Z}_{\geq 0}$, such that $\forall k \in \mathcal{V}$ and $\forall n \in \mathbb{Z}_{\geq 0}$, there exists $t \in [t_n, t_{n+1})$ such that $k \in \sigma(t)$. Then, the agents almost always converge to a set of equilibria corresponding to the critical points of V_{circ}. The only asymptotically stable equilibria are the local minima of V_{circ}.*

Proof idea

Under the assumptions of the proposition, denoting $\sum_{j=1}^{N} a_{jk} e^{i(\theta_j - \theta_k)} + \beta = \varrho_k e^{i u_k}$,

$$V_{\text{circ}}(t+1) - V_{\text{circ}}(t) = -2 \sum_{k \in \sigma(t)} (\varrho_k + \beta) \left(\sin\left(\frac{u_k}{2}\right)\right)^2 \leq 0.$$

Since every agent k is updated at an infinite number of time instants, u_k or ϱ_k (making u_k undefined) must go to 0 when t goes to $+\infty$, $\forall k \in \mathcal{V}$. The case $\varrho_k = 0$ has zero measure and can appear only "by chance" because it is the global maximum of V_{circ} with respect to θ_k. If u_k goes to 0, then agent k asymptotically approaches an equilibrium set; only minima can be asymptotically stable for a descent algorithm.

For synchronous update, it is again necessary to impose a bound on the motion of the agents. However, this bound is not easy to find. The following result provides a conservative bound on β.

Proposition 10.6

(Adapted from Ref. [26]) *Consider a set of N agents evolving on S^1 by applying (10.8) synchronously with fixed, undirected, and unweighted $\mathbb{G}$. Assume that*

$$\beta \geq d_{\max}\left(\frac{2}{M^*} + 1\right), \quad \text{where} \quad \frac{e^{M^*} - 1}{M^*} = 1 + \frac{d_{\max}}{d_{\text{sum}}}$$

with $d_{\text{sum}} = \sum_{j=k}^{N} d_k^{(i)}$ and $d_{\text{max}} = \max_{k \in \nu}(d_k^{(i)})$, where $d_k^{(i)}$ is the in-degree of agent k. Then, the agents converge to a set of equilibria corresponding to the critical points of V_{circ}. The only asymptotically stable equilibria are the local minima of V_{circ}.

Proof idea

The proof shows that $V_{\text{circ}}(t+1) - V_{\text{circ}}(t) \leq 0$ for synchronous operation and the bound on β. See Refs [20, 26] for complete computations.

A problem with the bound of Proposition 10.6 is that each agent must know d_{sum} and d_{max}, which is information about the (communication structure of the) whole swarm.

In the absence of inertia ($\beta = 0$), (10.8) can lead to a limit cycle in synchronous operation, at least for some $\mathbb{G}$ (see Ref. [20] for an example).

10.5.1.1 Hopfield Network

This model proposed in Ref. [27] considers N neurons with states $x_k \in \{-1, 1\}$. The discrete-time update law for the states of the neurons is

$$x_k(t+1) = \text{sign}\left(\sum_{j=1}^{N} a_{jk} x_j(t) + \xi_k\right), \quad 1, 2, \ldots, N, \tag{10.15}$$

where ξ_k is a firing threshold. Considering $V_H = \frac{-1}{2}\sum_{k=1}^{N}\sum_{j=1}^{N} a_{jk} x_j x_k - \sum_{k=1}^{N} x_k \xi_k$, Ref. [27] shows that when (10.15) is applied asynchronously with a random update sequence, the property $V_H(t+1) \leq V_H(t)$ always holds and the network eventually reaches a local minimum of V_H. In contrast, the system can go into a limit cycle under synchronous operation (see Ref. [28]).

Defining the sphere S^n of dimension n as $\{x_k \in \mathbb{R}^{n+1} : \|x_k\| = 1\}$, the set $\{-1, 1\}$ can be seen as "S^0", while the circle is S^1. For $\xi_k = 0$, (10.15) is in fact the strict analogue of (10.8) for "the sphere of dimension 0" – namely, moving toward the neighbors in the embedding vector space and projecting back to the state space. The absence of inertia in (10.15) would correspond to $\beta = 0$ in (10.8). Both (10.8) and (10.15) can be viewed as projections of descent algorithms for a symmetric quadratic potential, which remain descent algorithms under locally asynchronous update such that convergence is ensured. Both algorithms can fail to converge and run into a limit cycle in synchronous operation.

Propositions 10.4–10.6 say that the stable equilibria are the minima of V_{circ}. Unfortunately, depending on $\mathbb{G}$, there may be local minima different from synchronization.

10.5.1.2 Local Equilibria for the Undirected Ring

The following example is taken from Ref. [29]. Consider N agents interconnected according to an undirected, unweighted ring graph. Then, the critical points of V_{circ} satisfy $\sin(\theta_{j_a(k)} - \theta_k) + \sin(\theta_{j_b(k)} - \theta_k) = 0$, where $j_a(k)$ and $j_b(k)$ are the two neighbors of k in the ring graph. This requires positions of consecutive agents in the ring graph to differ either by θ_0 or by $\pi - \theta_0$, for some $\theta_0 \in [0, \pi/2]$ well chosen such that

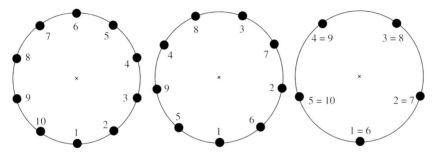

Figure 10.2 Several balanced configurations that are stable for the undirected ring graph; agents are numbered in the order of the ring, for example, agent 3 is connected to agents 2 and 4.

the sum of all angle differences is a multiple of 2π. Stability of these equilibria can be assessed by examining the Hessian of V_{circ}. This leads to the conclusion that each configuration with $|\theta_j - \theta_k| = \theta_0 < \pi/2 \ \forall (j,k) \in \mathcal{E}$ is locally asymptotically stable under (10.12) or (10.8), and all other configurations are unstable. Thus, in a stable configuration, consecutive agents in the ring graph are separated by θ_0 on the circle, for some $\theta_0 \in (-\pi/2, \pi/2)$ satisfying $N\theta_0 = 2a\pi$ with $a \in \mathbb{Z}$. The case $\theta_0 = 0$ corresponds to synchronization. When $N \geq 5$, stable configurations exist with $\theta_0 \neq 0$ (see Figure 10.2). For all these configurations, $\sum_{k=1}^{N} e^{i\theta_k} = 0$, therefore they are said to be *balanced*.

10.5.1.3 Stable Configurations Are Graph Dependent

It is an open question to characterize, with graph theoretic properties, which graphs admit no local minima of V_{circ} different from synchronization. The following result shows that, in fact, any configuration that is sufficiently "spread" on the circle is stable under the synchronization algorithms for a well-chosen weighted digraph.

Proposition 10.7

Consider a set of N agents distributed on S^1 in a configuration $\{\theta_k\}$ such that for every k, there is at least one agent located in $(\theta_k, \theta_k + \pi/2)$ and one located in $(\theta_k - \pi/2, \theta_k)$; such a configuration requires $N \geq 5$. Then, there exists a positively weighted and strongly connected δ-digraph making this configuration locally exponentially stable under (10.12) with $\alpha > 0$.

Proof idea

Choose nonzero weights a_{jk} only for the $\theta_j \in (\theta_k - \pi/2, \theta_k + \pi/2)$, and such that $r_k \sum_{j=1}^{N} a_{jk}(x_j - x_k)$ is aligned with $x_k = e^{i\theta_k}$.

For any of the weight choices that locally stabilize specific configurations, synchronization is also exponentially stable, but only locally. The equilibrium configurations of (10.8) and (10.12) different from synchronization are formalized in Ref. [21] as consensus configurations. The same paper also considers the related

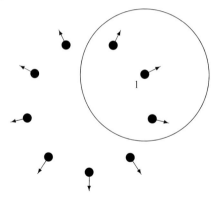

Figure 10.3 Initial conditions for divergent behavior in Vicsek model. The black disks denote agent positions, the arrows denote orientations, and the circle represents the sensing region of agent 1.

problem of "spreading" agents on the circle, which is formalized with the notions of *anticonsensus* and *balancing* configurations.

10.5.1.4 Structurally Stable Divergent Behavior in Vicsek Model

The above example of stable equilibria for the undirected ring allows to illustrate a situation where Vicsek model diverges. Consider $N \geq 5$ agents initiated as in Figure 10.3: (i) initial positions $r_k(0) \in \mathbb{R}^2$ are regularly distributed on a circle such that each agent can sense only its immediate neighbor on the left and on the right; (ii) initial orientations $\theta_k(0) \in S^1$ point radially outward of the circle formed by the positions. Then, the update equation for agent orientations $\theta_k(t)$ is exactly in a stable configuration different from synchronization under a ring interconnection graph. The agents move radially outward; at a particular time step, all communication links drop. Stability of the equilibrium for the orientations and the fact that all communication links still disconnect at the same instant when positions are slightly shifted ensures that the divergent behavior is observed in an open neighborhood of initial conditions around this ideal situation. Examining stable equilibria of (10.8) and (10.12) for $\mathbb{G}$ different from an undirected ring, one sees that the stable divergent behavior remains if the sensing regions are increased such that each agent initially has several neighbors on the left and on the right.

10.5.2
Limit Sets Different from Equilibrium

Section 10.5.1 lists cases where (10.12) or (10.8) do not converge to *synchronization*, but still to a set of equilibria. There are also cases where the agents do not converge to a set of equilibria.

For fixed undirected graphs, the swarm is ensured to converge to a set of equilibria, except for the discrete-time algorithm when β is too small (see Proposition 10.6). In the latter case, behavior of (10.8) is not as clear and the system may run into a limit cycle; see Refs [20, 26] for details.

Periodic and quasi-periodic behaviors can be easily constructed for (10.12) with fixed *directed* graphs.

The simplest such behavior is called *cyclic pursuit*: each agent k is attracted by its neighbors to move (say) clockwise, and the agents keep turning without synchronizing. A classical situation of stable cyclic pursuit is a directed ring graph with consecutive agents separated by $2\pi/N$.

In basic cyclic pursuit, agents keep moving on the circle but relative positions remain constant. A more meaningful periodic behavior occurs when *relative positions periodically vary* in time. Such situations can be built with agents partitioned into two sets such that each set is in cyclic pursuit at a different velocity. Start, for instance, with two unweighted directed ring graphs of N_1 and N_2 agents, where $N_1 + N_2 = N$ and $N_1 \neq N_2$; consecutive agents in each ring are separated by $2\pi/N_1$ and $2\pi/N_2$. Then, the resulting behavior is satisfactory, but the overall graph is not connected. To obtain a strongly connected graph, each agent of the first ring can be coupled to all the agents of the second ring and conversely; indeed, for a set of regularly spaced agents, $\sum_k e^{i\theta_k} = 0$, so coupling an agent to such a set does not change its behavior.

Likewise, a *quasi-periodic variation* of relative positions is obtained when several sets of agents move in cyclic pursuit with irrational velocity ratios. This can be built, for instance, with unitary graph weights and $\alpha = 1$ if one set has x agents in a splay state for an undirected ring graph $\Leftrightarrow d\theta_k/dt = 0$, the second set has 6 agents in cyclic pursuit with a directed ring graph $\Leftrightarrow d\theta_k/dt = \sqrt{3}$, and the third set has 12 agents in cyclic pursuit with a directed ring graph $\Leftrightarrow d\theta_k/dt = 1$.

Finally, an example of *disorderly looking quasi-periodic motion* can be built by adding to the previous situation an agent that is influenced by one agent in each of the three rings; the motion of this agent is illustrated on Figure 10.4.

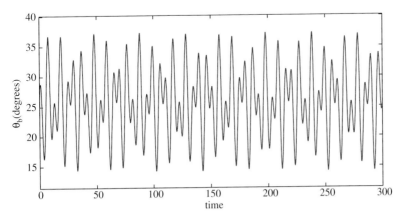

Figure 10.4 Illustration of disorderly looking quasi-periodic motion of an agent k for which $d\theta_k/dt = \sum_{j=1}^{3} \sin(\theta_j - \theta_k)$, where θ_1 belongs to a regularly spaced undirected ring, $d\theta_1/dt = 0$; θ_2 is in cyclic pursuit with 5 other agents, $d\theta_2/dt = \sqrt{3}$; θ_3 is in cyclic pursuit with 11 other agents, $d\theta_3/dt = 1$.

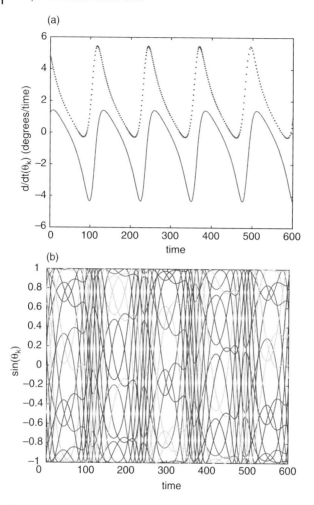

Figure 10.5 Motion of agents applying (10.12) with fixed directed coupling, such that they all periodically revert their direction of motion. Set A has nine agents regularly spaced by $2\pi/9$ on the circle. Set B has nine agents also regularly spaced, but initially rotated by $\pi/18$ with respect to A. In A, $d\theta_k/dt = 0.04 \sin(\theta_j-\theta_k) + 0.05 \sin(\theta_l-\theta_k)$, where j is the agent of A for which $\theta_j-\theta_k = -2\pi/9$, and l is the agent of B for which initially $\theta_l-\theta_k = 7\pi/18$. In B, $d\theta_k/dt = 0.07 \sin(\theta_j-\theta_k) + 0.05 \sin(\theta_l-\theta_k)$, where j is the agent of B for which $\theta_j-\theta_k = 2\pi/9$ and l is the agent of A for which initially $\theta_l-\theta_k = 5\pi/18$. (a) velocities of the two sets (continuous curve for set A and dotted curve for set B). (b) evolution of $\sin(\theta_k)$ for all agents k.

It is possible to build situations of fixed directed coupling with even more surprising behavior. Figure 10.5 represents the motion of two sets of agents in cyclic pursuit, with coupling among agents of the two sets and initial positions such that all the agents periodically revert their direction of motion.

For time-varying graphs, the situation is even more complicated. Since many different configurations can be stable on the circle depending on $\mathbb{G}$, the swarm can be driven toward different equilibria during longer or shorter time spans, implying no particular characterization of the swarm's behavior if $\mathbb{G}(t)$ can be arbitrary. In practice, synchronization is often eventually observed. This is because semicircle, they can only converge to synchronization. as soon as all agents lie in the same. But other asymptotic behaviors are also possible.

The diversified, poorly characterized behavior of (10.8) or (10.12) with directed and time-varying $\mathbb{G}$ is in strong contrast with the behavior of the consensus algorithm on vector space, which is fully characterized by Proposition 10.2. In addition, Proposition 10.2 can be extended to the case where time delays are present along the communication links (see Ref. [16]), while the behavior of (10.8) or (10.12) under time delays is still under investigation even for fixed undirected $\mathbb{G}$. Even for the complete graph, delays may lead to stable synchronized solutions, stable "spread" solutions, and periodic oscillations [30].

Finally, it must be noted that the graph modeling interagent communication often depends on the states of the agents, like, for instance, in full Vicsek model [4]. Studying the interaction of state-dependent graphs with algorithms that are not specifically designed for particular graph behavior is a difficult problem, which goes beyond the scope of the present paper.

10.6
Algorithms for Global Synchronization

Section 10.5 highlights that consensus algorithms on the circle may exhibit complicated behaviors. In a control framework, a natural question is whether the update rules (10.8) and (10.12) can be modified to enforce better synchronization properties.

10.6.1
Modified Coupling for Fixed Undirected Graphs

For fixed undirected graphs, algorithms (10.8) and (10.12) guarantee convergence to an equilibrium set, but stable equilibria different from synchronization may exist. Such spurious equilibria can be rendered unstable by reshaping the way agents are attracted toward their neighbors. A continuous-time setting is chosen for convenience; a similar result could be developed in discrete time. For simplicity, $\mathbb{G}$ is assumed unweighted.

Consider a continuous-time synchronization algorithm on S^1 of the form

$$\frac{d}{dt}\theta_k = \sum_{j \to k} f(\theta_j - \theta_k), \quad k = 1, 2, \ldots, N. \tag{10.16}$$

The function $f : S^1 \to \mathbb{R}$ is called the *coupling function*. For (10.12), $f(\theta) = \sin(\theta)$. Section 10.5.1 examines local equilibria of (10.12) for the undirected ring graph and

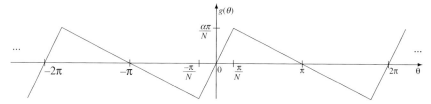

Figure 10.6 Alternative coupling function g.

concludes that a configuration is stable if interconnected agents are closer than $\pi/2$ (because $df(\theta)/d\theta > 0$ for $|\theta| < \pi/2$) and unstable if they are further apart than $\pi/2$ (because $df(\theta)/d\theta < 0$ for $|\theta| > \pi/2$). If $f(\theta)$ is modified to have a positive slope only up to π/a for some $a > 2$, then connected agents must be closer than π/a at a stable equilibrium for the ring graph; taking $a/N > 1/2$, it becomes impossible to distribute the agents as in Figure 10.2 and synchronization is the only stable equilibrium. This motivates the following.

Assume (a bound on) the number N of agents in the swarm is available. Define

$$g(\theta) = \begin{cases} \dfrac{-a}{N-1}(\pi+\theta) & \text{for } \theta \in \left[-\pi, -\dfrac{\pi}{N}\right] \\ a\theta & \text{for } \theta \in \left[-\dfrac{\pi}{N}, \dfrac{\pi}{N}\right] \\ \dfrac{a}{N-1}(\pi-\theta) & \text{for } \theta \in \left[\dfrac{\pi}{N}, \pi\right] \end{cases} \qquad (10.17)$$

for some $a > 0$, extended 2π periodically outside the above intervals, as represented in Figure 10.6. Function $g(\theta)$ is the gradient of (with 2π-periodic extension)

$$(z(\theta))^2 = \begin{cases} \dfrac{a\pi^2}{2N(N-1)} + \dfrac{a}{N-1}\left(-\pi\theta - \dfrac{\theta^2}{2}\right) & \text{for } \theta \in \left[-\pi, -\dfrac{\pi}{N}\right] \\ \dfrac{a}{2}\theta^2 & \text{for } \theta \in \left[-\dfrac{\pi}{N}, \dfrac{\pi}{N}\right] \\ \dfrac{a\pi^2}{2N(N-1)} + \dfrac{a}{N-1}\left(\pi\theta - \dfrac{\theta^2}{2}\right) & \text{for } \theta \in \left[\dfrac{\pi}{N}, \pi\right], \end{cases}$$

which is even, has a minimum for $\theta = 0$, a maximum for $\theta = \pi$, and evolves monotonically and continuously in between, similar to the sinusoidal distance measure $(2\sin(\theta/2))^2$.

Choosing $f(\theta) = g(\theta)$ in (10.16) defines a synchronization algorithm that satisfies all invariance and communication constraints and whose only stable configuration is synchronization for fixed undirected $\mathbb{G}$. Rigorously, the edges in $g(\theta)$ should be smoothed to make it continuously differentiable everywhere; this changes nothing to the general argument.

Proposition 10.8

Consider a swarm of N agents, interconnected according to a connected fixed undirected graph $\mathbb{G}$, which evolve on S^1 by applying (10.16) with $f(\theta) = g(\theta)$ defined by (a smoothed version of) (10.17). The agents always converge to the set of equilibria corresponding to the critical points of $V_g = 1/2 \sum_k \sum_{j \rightsquigarrow k} (z(\theta_j - \theta_k))^2$. Moreover, the only asymptotically stable equilibrium is synchronization.

Proof idea

(See Ref. [20] for a full proof.) The agents always converge to a set of critical points of V_g because the algorithm is a gradient descent for V_g. Synchronization, as the global minimum of V_g, is stable. Stability of other equilibria is characterized by examining the Hessian of V_g. It is shown that for the interaction function $g(\theta)$, if the Hessian is positive semidefinite with 0 eigenvalue only in the direction of uniform motion ($\theta_k \to \theta_k + a \ \forall k \in \mathcal{V}$), then the graph $\mathbb{G}_p$, containing edge $\{j, k\}$ if and only if j and k are closer than π/N, must be connected. Then, all agents must be within a semicircle and convexity arguments impose synchronization like on the real line.

10.6.2
Introducing Randomness in Link Selection

The modified coupling function solves the problem of spurious local minima in V_{circ}. However, it may introduce numerous unstable equilibria. Moreover, for varying and directed graphs, the behavior of (10.16) is not better characterized than for (10.12). This section introduces the so-called "gossip algorithm" (see Ref. [31] and references therein) in order to improve synchronization behavior on the circle. Thanks to the introduction of randomness, it achieves *global asymptotic synchronization with probability* 1 for directed and time-varying $\mathbb{G}$. It is described in discrete time for easier formulation.

The nice convergence properties of (10.8) and (10.12) when $\mathbb{G}$ is a tree motivate to keep the update law (10.8), but at each time select at most one of the in-neighbors in $\mathbb{G}(t)$ for each agent. In order to satisfy the equivalence of all agents, k may not privilege any of its neighbors – it is just allowed, for weighted $\mathbb{G}$, to take the different weights of the corresponding edges into account. Always choosing the neighbor with maximum weight could disconnect the swarm. Therefore, it is necessary to select the retained neighbor *randomly* among the $j \rightsquigarrow k$. A natural probability distribution for neighbor selection would follow the weights of the edges, but in theory any distribution with nonzero weight on each link is admissible. With the proposed edge selection procedure, the neighbor chosen at time $t+1$ is independent of the neighbor chosen at time t (up to, for varying graphs, a possible dependence of $\mathbb{G}(t+1)$ on $\mathbb{G}(t)$).

10.6.2.1 Gossip Algorithm (Directed)

At each update t,

1) each agent k randomly selects an agent $j \to k$ with probability $a_{jk}/(\beta + \sum_{l \to k} a_{lk})$, where $\beta > 0$ is the weight for choosing no agent;
2) $\theta_k(t+1) = \theta_j(t)$ if agent k chooses neighbor j at time t and $\theta_k(t+1) = \theta_k(t)$ if it chooses no neighbor.

A variant of the gossip algorithm exists in which the random graph is *undirected* at each time step. In this variant, agents have to select each other in order to move by averaging their positions. The advantages of this variant on S^1 are not clear, and proving global asymptotic synchronization with probability 1 is somewhat more difficult; see Ref. [32] for details.

The directed gossip algorithm proposed above is extreme in the sense that agent k directly jumps to the position of its selected neighbor. A more moderate directed gossip algorithm would apply the update law $\theta_k(t+1) = \arg(\alpha\, e^{i\theta_k(t)} + e^{i\theta_j(t)})$ with $\alpha > 0$.

The authors of Ref. [31] perform a detailed analysis of a gossip algorithm for synchronization in vector spaces. Convergence toward synchronization is always ensured on vector spaces and the problem is to quantify the convergence *rate* as a function of $\mathbb{G}$ and probability (i.e., weights) distribution. On the circle, convergence toward synchronization is not obvious *a priori* but, in fact, it holds under the same assumptions.

Definition 10.3

N agents asymptotically synchronize with probability 1 if for any initial condition, for any $\varepsilon > 0$ and $\kappa \in (0, 1)$, there exists a time T after which the maximal distance $|\theta_k(T) - \theta_j(T)|$ between any pair of agents is smaller than ε with probability larger than κ.

Proposition 10.9

Consider a set of N agents interconnected according to a uniformly connected δ-digraph $\mathbb{G}$. If the agents apply the (directed) gossip algorithm with a fixed finite $\beta > 0$, then they asymptotically synchronize with probability 1.

Proof idea

(See Ref. [32] for a full proof.) Consider a set of links forming a directed tree of root k. For $\mathbb{G}(t)$ uniformly connected, there exists a sequence of link choices, implementable on any time interval $[t, t + T_s]$ for some finite T_s, which builds this tree (and only this tree, potentially selecting no move for many time instants!) sequentially from the root to its leaves. This sequence synchronizes the agents at the position of the root *for any initial conditions.* Moreover, its finite length implies a finite (though potentially tiny) probability to be selected during any time interval of length T_s. Therefore, when t goes to $+\infty$, the probability that this sequence never appears goes to 0. Since it suffices that the sequence appears once to ensure synchronization, this concludes the proof.

The convergence proof of Proposition 10.9 can be adapted for the moderate version of the directed gossip algorithm, with inertia $\alpha > 0$, as described in Ref. [20].

Under the initial gossip algorithm (inertia-less, i.e., with $\alpha = 0$), agents, in fact, jump between a discrete set of possible positions corresponding to the initial positions of the N agents. This highlights that the directed gossip algorithm can, in fact, be applied on *any set of symbols*. Proposition 10.9 purely relies on the evolution of agents between N different "symbols," *completely independent of the underlying manifold.* Every time a position is left empty (implying that the synchronization process progresses as an agent joins other ones), that position can never be reached again in the future; this process goes on until all agents are on the same position after a finite time.

In this context, a natural measure of convergence rate is the *expected synchronization time*, that is, the average time, over all possible link choices, after which all agents are on the same position. A Markov chain framework can be applied to obtain an explicit formula for the expected synchronization time, at least for fixed graph $\mathbb{G}$ (see Ref. [20]); unfortunately, the complexity of the explicit formula exponentially grows with the number of agents, and it seems difficult to extract the influence of the link choice probability distribution. The expected synchronization time – including its numerical value – is independent of S^1 and *independent of the initial positions of the agents* (unless some agents are initially perfectly synchronized). The only remaining parameters are the graph $\mathbb{G}$ and the probability distribution. There seems to be an interesting interplay between these two parameters; it is therefore challenging for the agents to optimize their convergence rate on the basis of their local information.

Simulations confirm that the gossip algorithm favors global synchronization. However, they also highlight that the probabilistic setting can lead to unnecessarily slow convergence rates: often a set of agents partitioned into two groups, located at θ_a and θ_b, respectively, keep oscillating between these two positions for an appreciable time before they synchronize on one of them.

10.6.3
Algorithms Using Auxiliary Variables

In the synchronization framework, the nonconvexity of S^1 can be "cheated" if the agents are able to communicate *auxiliary variables* in addition to their positions on the circle. A different viewpoint on this procedure is that the limited number of communication links for information flow is compensated by sending larger communication packets along existing links. Such strategies allow to recover the synchronization properties of vector spaces for almost all initial conditions. Their potential interest lies more in engineering applications than in physical modeling, where communication of auxiliary variables is questionable. In the present paper, the relevant algorithms are just briefly mentioned for completeness.

The first use of auxiliary variables is to build a reduced communication network, in which a leader is identified that then attracts all the agents. The leader election/spanning tree construction would be completely independent of the agents' motion on S^1 and should be achieved after a finite time; then, applying a synchronization algorithm with this leader/spanning tree would ensure synchronization. See Ref. [33] for a distributed algorithm that achieves this preliminary network construction. Note that for termination of the network construction after finite time, (a bound on) the total number N of agents must be available to each agent. Also, it is not clear how well the spanning tree construction can be adapted to time-varying graphs.

The second use of auxiliary variables is to reach consensus on a *reference synchronization point*: an auxiliary variable in $\mathbb{R}^2$ is associated with each agent and a consensus algorithm is run on these auxiliary variables. Each agent individually tracks the projection of its auxiliary variable on S^1. Thanks to the fact that S^1 is equivalent to the Lie group $SO(2)$, this can be implemented in a way that satisfies the coupling symmetry hypotheses. For uniformly connected $\mathbb{G}$, the consensus algorithm on auxiliary variables defines a point in the plane, and synchronization is ensured for almost all initial conditions. See Ref. [34] for details and a formal proof.

10.7
Generalizations on Compact Homogeneous Manifolds

Although the discussion in this chapter focuses on the circle for simplicity, it is representative of more general manifolds. Compact homogeneous manifolds are manifolds that can be viewed as the quotient of a compact Lie group by one of its subgroups. They include the n-dimensional sphere S^n, all compact Lie groups like, for example, the group of rotations $SO(n)$, and the Grassmann manifolds $Grass(p, n)$.

In Ref. [21], the developments of Section 10.3 are extended to compact homogeneous manifolds $\mathcal{M}$ that are embedded in $\mathbb{R}^m$ such that the Euclidean norm $\|x\|$ in $\mathbb{R}^m$ is constant over $x \in \mathcal{M}$. The cost function and gradient algorithms are generalized, and an interpretation in terms of moving toward an appropriately defined *average* of positions on $\mathcal{M}$ is given. In particular, the cost function measures the Euclidean distance between agents *in the embedding space*, also called the *chordal distance*. Convergence properties are analyzed, and stable local equilibria are formalized as *consensus* configurations. An "opposite" algorithm, in which agents move away from their neighbors to spread on the circle, is proposed along the same lines and its convergence properties are formalized with anticonsensus and balancing configurations.

Regarding global synchronization, the results of Section 10.4 both for trees and complete graphs and for agents initially located within convex sets remain valid on more general manifolds. Obstacles like those illustrated in Section 10.5 appear as well. The modified coupling algorithm of Section 10.6.1 should, in principle, have an extension on general manifolds, but the current convergence proof is algebraic rather than geometric and cannot be repeated. The (directed) gossip algorithm of Section 10.6.2 works on any manifold – in fact, any set; the undirected variant is less easily generalized. The algorithm mentioned in Section 10.6.3, specifying a "reference synchronization state" with auxiliary variables, is generalized to compact homogeneous manifolds in Ref. [21] (modulo the fact that a meaningful communication of auxiliary variables sometimes requires a common external reference frame). The same type of algorithm with auxiliary variables can also be used for balancing on the circle and on compact homogeneous manifolds.

10.8
Conclusions

This chapter proposes a natural extension of the "consensus" framework, where agents have to agree on a common value, from values in *vector spaces* to values on the *circle*. This leads to a related interpretation of Kuramoto and Vicsek models of the literature.

Although convergence is similar to vector spaces in some specific situations, in general the behavior of consensus algorithms on the circle is much more diversified, allowing local equilibria, limit cycles, and essentially any type of behavior when the interconnection graph can vary freely.

Global synchronization properties can be recovered on the circle (i) by modifying the coupling between agents (Section 10.6.1), (ii) in a stochastic setting where at most one neighbor is randomly selected at each time step (Section 10.6.2), and (iii) less surprisingly, by assisting the agreement process with auxiliary variables in a vector space (Section 10.6.3).

Several questions remain open to characterize and optimize the *global* behavior of these simple consensus algorithms on the circle; this indicates possible directions of interest for future research.

Acknowledgments

This chapter presents research results of the Belgian Network DYSCO (Dynamical Systems, Control, and Optimization), funded by the Interuniversity Attraction Poles Programme, initiated by the Belgian State, Science Policy Office. The scientific responsibility rests with its authors. Professors L. Scardovi, J. Hendrickx, V. Blondel, E. Tuna, P-A. Absil, and N. Leonard are acknowledged for interesting discussions related to this subject. The second author is supported as an FNRS fellow (Belgian Fund for Scientific Research).

Appendix 10.A: Notions of Graph Theory

In the framework of coordination with limited interconnections between agents, it is customary to represent communication links by means of a *graph* (see, for instance, Refs [35, 36]).

Definition 10.A.1

A directed graph $\mathbb{G}(\mathcal{V}, \mathcal{E})$ (short digraph $\mathbb{G}$) is composed of a finite set $\mathcal{V}$ of vertices, and a set $\mathcal{E}$ of edges that represent interconnections among the vertices as ordered pairs (j, k) with j and $k \in \mathcal{V}$.

A weighted digraph $\mathbb{G}(\mathcal{V}, \mathcal{E}, \mathcal{A})$ is a digraph associated with a set $\mathcal{A}$ that assigns a positive weight $a_{jk} \in \mathbb{R}_{>0}$ to each edge $(j, k) \in \mathcal{E}$.

An unweighted graph is often considered a weighted graph with unit weights. A weighted graph can be defined only by its vertices and weights, by extending the weight set to all pairs of vertices, and by imposing $a_{jk} = 0$ if and only if (j, k) does not belong to the edges of $\mathbb{G}$. A digraph is said to be *undirected* if $a_{jk} = a_{kj} \,\forall j, k \in \mathcal{V}$. It may happen that $(j, k) \in \mathcal{E}$ whenever $(k, j) \in \mathcal{E} \,\forall j, k \in \mathcal{V}$, but $a_{jk} \neq a_{kj}$ for some $j, k \in \mathcal{V}$; in this case, the graph is called *bidirectional*. Equivalently, an unweighted undirected graph can be defined as a set of vertices and a set of *unordered* pairs of vertices.

In this chapter, each agent is identified with a vertex of a graph; the N agents = vertices are designed by positive integers $1, 2, \ldots, N$, so $\mathcal{V} = \{1, 2, \ldots, N\}$. The presence of edge (j, k) has the meaning that agent j sends information to agent k, or equivalently, agent k measures quantities concerning agent j. It is assumed that no "communication link" is needed for an agent k to get information about itself, so $\mathbb{G}$ contains no self-loops: $(k, k) \notin \mathcal{E} \,\forall k \in \mathcal{V}$. In visual representations of a graph, a vertex is depicted by a point and edge (j, k) by an arrow from j to k. Therefore, a frequent alternative notation for $(j, k) \in \mathcal{E}$ is $j \rightsquigarrow k$. One also says that j is an *in-neighbor* of k and k is an *out-neighbor* of j. In the visual representation of an undirected graph, all arrows are bidirectional; therefore, arrowheads are usually dropped. One simply says that j and k are *neighbors* and writes $j \sim k$ instead of $j \rightsquigarrow k$ and $k \rightsquigarrow j$. The *in-degree* of vertex k is $d_k^{(i)} = \sum_{j=1}^{N} a_{jk}$. The *out-degree* of vertex k is $d_k^{(o)} = \sum_{j=1}^{N} a_{kj}$. A digraph is said to be *balanced* if $d_k^{(i)} = d_k^{(o)} \,\forall k \in \mathcal{V}$; in particular, undirected graphs are balanced.

The *adjacency matrix* $A \in \mathbb{R}^{N \times N}$ of a graph $\mathbb{G}$ contains a_{jk} in row j, column k; it is symmetric if and only if $\mathbb{G}$ is undirected. Denote by $|\mathcal{E}|$ the number of edges in $\mathbb{G}$. For a digraph $\mathbb{G}$, each column of the *incidence matrix* $B \in (\{-1, 0, 1\})^{N \times |\mathcal{E}|}$ corresponds to one edge and each row to one vertex; if column m corresponds to edge (j, k), then

$$b_{jm} = -1, \quad b_{km} = 1, \quad \text{and} \quad b_{lm} = 0 \text{ for } l \notin \{j, k\}.$$

For an undirected graph $\mathbb{G}$, each column corresponds to an undirected edge; an *arbitrary orientation* (j, k) or (k, j) is chosen for each edge and B is built for the resulting directed graph. Thus, B is not unique for a given $\mathbb{G}$, but $\mathbb{G}$ is unique for a given B.

The in- and out-degrees of vertices $1, 2, \ldots, N$ can be assembled in diagonal matrices $D^{(o)}$ and $D^{(i)}$. The *in-Laplacian* of $\mathbb{G}$ is $L^{(i)} = D^{(i)} - A$. Similarly, the associated *out-Laplacian* is $L^{(o)} = D^{(o)} - A$. For a balanced graph $\mathbb{G}$, the *Laplacian* $L = L^{(i)} = L^{(o)}$. The standard definition of Laplacian L is for undirected graphs. For the latter, L is symmetric and, remarkably, $L = BB^T$. For general digraphs, by construction, $(\mathbf{1}_N)^T L^{(i)} = 0$ and $L^{(o)} \mathbf{1}_N = 0$, where $\mathbf{1}_N$ is the column vector of N ones. The spectrum of the Laplacian reflects several interesting properties of the associated graph, especially in the case of undirected graphs (see for example Ref. [35]). In particular, it reflects its *connectivity* properties.

A *directed path* of length l from vertex j to vertex k is a sequence of vertices $v_0, v_1, \ldots, v_l$ with $v_0 = j$ and $v_l = k$ and such that $(v_m, v_{m+1}) \in \mathcal{E}$ for $m = 0, 1, \ldots, l-1$. An *undirected path* between vertices j and k is a sequence of vertices $v_0, v_1, \ldots, v_l$ with $v_0 = j$ and $v_l = k$ and such that $(v_m, v_{m+1}) \in \mathcal{E}$ or $(v_{m+1}, v_m) \in \mathcal{E}$, for $m = 0, 1, \ldots, l-1$. A digraph $\mathbb{G}$ is *strongly connected* if it contains a directed path from every vertex to every other vertex (and thus also back to itself). A digraph $\mathbb{G}$ is *root connected* if it contains a node k, called the root, from which there is a path to every other vertex (but not necessarily back to itself). A digraph $\mathbb{G}$ is *weakly connected* if it contains an undirected path between any two of its vertices. For an undirected graph $\mathbb{G}$, all these notions become equivalent and are simply summarized by the term *connected*. For $\mathbb{G}$ representing interconnections in a network of agents, clearly coordination can take place only if $\mathbb{G}$ is connected. If this is not the case, coordination will be achievable only separately in each connected component of $\mathbb{G}$. A more interesting discussion of connectivity arises when the graph $\mathbb{G}$ can vary with time. Before discussing this case, the following summarizes some spectral properties of the Laplacian that are linked to the connectivity of the associated graph.

Properties (Laplacian): Consider the out-Laplacian $L^{(o)}$ of digraph $\mathbb{G}$.
a) All eigenvalues of $L^{(o)}$ have nonnegative real parts.
b) If $\mathbb{G}$ is strongly connected, then 0 is a simple eigenvalue of $L^{(o)}$.
c) Expression $x^T L x$, with $x \in \mathbb{R}^N$, is positive semidefinite if and only if $\mathbb{G}$ is balanced.

If $\mathbb{G}$ is undirected, the Laplacian L has the following properties:
d) L is symmetric positive semidefinite.
e) The algebraic and geometric multiplicity of 0 as an eigenvalue of L is equal to the number of connected components in $\mathbb{G}$.

In a coordination problem, interconnections among agents can vary with time, as some links are dropped and others are established. In this case, the communication links are represented by a *time-varying graph* $\mathbb{G}(t)$ in which the vertex set $\mathcal{V}$ is fixed (by convention), but edges $\mathcal{E}$ and weights $\mathcal{A}$ can depend on time. All the previous definitions carry over to time-varying graphs; simply, each quantity depends on time. To prevent edges from vanishing or growing indefinitely, this chapter considers δ-*digraphs*, for which the elements of $\mathcal{A}(t)$ are bounded and satisfy the threshold $a_{jk}(t) \geq \delta > 0 \, \forall (j, k) \in \mathcal{E}(t)$, for all t. In addition, in continuous time $\mathbb{G}$ is assumed to be piecewise continuous. For δ-digraphs $\mathbb{G}(t)$, it is intuitively clear that coordination may be achieved if information exchange is "sufficiently frequent," without requiring it to take place all the time. The following definition of "integrated connectivity over time" can be found in Refs [12–14, 18].

Definition 10.A.2

(Adapted from Refs [13, 14].) In discrete time, for a δ-digraph $\mathbb{G}(\mathcal{V}, \mathcal{E}(t), \mathcal{A}(t))$ and some constant $T \in \mathbb{Z}_{\geq 0}$, define the graph $\bar{\mathbb{G}}(\mathcal{V}, \bar{\mathcal{E}}(t), \bar{\mathcal{A}}(t))$, where $\bar{\mathcal{E}}(t)$ contains all edges that appear in $\mathbb{G}(\tau)$ for $\tau \in [t, t+T]$ and $\bar{a}_{jk}(t) = \sum_{\tau=t}^{t+T} a_{jk}(\tau)$. Similarly, in continuous time, for a δ-digraph $\mathbb{G}(\mathcal{V}, \mathcal{E}(t), \mathcal{A}(t))$ and some constant $T \in \mathbb{R}_{>0}$, define the graph $\bar{\mathbb{G}}(\mathcal{V}, \bar{\mathcal{E}}(t), \bar{\mathcal{A}}(t))$ by

$$\bar{a}_{jk}(t) = \begin{cases} \int_t^{t+T} a_{jk}(\tau)d\tau, & \text{if } \int_t^{t+T} a_{jk}(\tau)d\tau \geq \delta \\ 0, & \text{if } \int_t^{t+T} a_{jk}(\tau)d\tau < \delta. \end{cases}$$

$(j,k) \in \bar{\mathcal{E}}(t)$,   if and only if $\bar{a}_{jk}(t) \neq 0$.

Then, $\mathbb{G}(t)$ is said to be uniformly connected over T if there exists a time horizon T and a vertex $k \in \mathcal{V}$ such that $\bar{\mathbb{G}}(t)$ is root connected with root k for all t.

The following graphs are regularly used in the present dissertation.

- The (equally weighted) *complete graph* is an unweighted, undirected graph that contains an edge between every pair of vertices.
- An *undirected ring* or *cycle graph* on $N > 1$ vertices is equivalent to an undirected path containing all vertices, to which is added an edge between the extreme vertices of the path. Similarly, a *directed ring* or *cycle graph* on $N > 1$ vertices is equivalent to a directed path containing all vertices, to which is added an edge from the last to the first vertex in the path.
- An *undirected tree* is a connected undirected graph in which it is impossible to select a subset of at least three vertices and a subset of edges among them to form an undirected cycle. A *directed tree* of root k is a root-connected digraph of root k, in which every vertex can be reached from k by following one and only one directed path. A (undirected) tree on N vertices has N-1 (undirected) edges.

In a directed tree $\mathbb{G}$, the (unique) in-neighbor of a vertex j is called its *parent* and its out-neighbors are its *children*. The root has no parent, and the vertices with no children are called the *leaves*. This can be carried over to an undirected graph after selecting an arbitrary root.

References

1 Lorentz, E.N. (1963) Deterministic nonperiodic flow. *J. Atmos. Sci.*, **20**, 130–141.

2 Setti, G., Mazzini, G., Rovatti, R., and Callegari, S. (2002) Statistical modeling of discrete-time chaotic processes: basic finite-dimensional tools and applications. *Proc. IEEE*, **90** (5), 662–690.

3 Strogatz, S.H. (2003) *Sync: The Emerging Science of Spontaneous Order*, Hyperion.

4 Vicsek, T., Czirók, A., Ben-Jacob, E., Cohen, I., and Shochet, O. (1995) Novel type of phase transition in a system of self-driven particles. *Phys. Rev. Lett.*, **75** (6), 1226–1229.

5 Pecora, L.M. and Carroll, T.L. (1990) Synchronization in chaotic systems. *Phys. Rev. Lett.*, **64**, 821–824.

6 Cortés, J., Martínez, S., and Bullo, F. (2004) Coordinated deployment of mobile sensing networks with limited-range interactions. Proceedings of the 43rd IEEE Conference on Decision and Control, pp. 1944–1949.

7 Leonard, N.E., Paley, D., Lekien, F., Sepulchre, R., Frantantoni, D., and Davis, R. (2007) Collective motion, sensor networks and ocean sampling. *Proc. IEEE*, **95** (1), 48–74.

8 Beugnon, C., Buvat, E., Kersten, M., and Boulade, S. (2005) GNC design for the DARWIN spaceborne interferometer. Proceedings of the 6th ESA Conference on Guidance, Navigation and Control Systems.

9 Monnier, J.D. (2003) Optical interferometry in astronomy. *Rep. Prog. Phys.*, **66**, 789–857.

10 Cucker, F. and Smale, S. (2007) Emergent behavior in flocks. *IEEE Trans. Automat. Contr.*, **52** (5), 852–862.

11 Kuramoto, Y. (1975) Self-entrainment of population of coupled nonlinear oscillators, in *International Symposium on Mathematical Problems in Theoretical Physics*, vol. 39 (ed. by H. Araki), Lecture Notes in Physics, Springer, p. 420.

12 Blondel, V.D., Hendrickx, J.M., Olshevsky, A., and Tsitsiklis, J.N. (2005) Convergence in multiagent coordination, consensus and flocking. Proceedings of the 44th IEEE Conference on Decision and Control.

13 Moreau, L. (2004) Stability of continuous-time distributed consensus algorithms. Proceedings of the 43rd IEEE Conference on Decision and Control, pp. 3998–4003.

14 Moreau, L. (2005) Stability of multi-agent systems with time-dependent communication links. *IEEE Trans. Automat. Contr.*, **50** (2), 169–182.

15 Olfati-Saber, R., Fax, J.A., and Murray, R.M. (2007) Consensus and cooperation in networked multi-agent systems. *Proc. IEEE*, **95** (1), 215–233.

16 Olfati-Saber, R. and Murray, R.M. (2004) Consensus problems in networks of agents with switching topology and time delays. *IEEE Trans. Automat. Contr.*, **49** (9), 1520–1533.

17 Olshevsky, A. and Tsitsiklis, J.N. (2006) Convergence rates in distributed consensus and averaging. Proceedings of the 45th IEEE Conference on Decision and Control, pp. 3387–3392.

18 Tsitsiklis, J.N. and Athans, M. (advisor) (1984) Problems in decentralized decision making and computation. PhD Thesis, MIT.

19 Sarlette, A., Bonnabel, S., and Sepulchre, R. (2010) Coordinated motion design on Lie groups. *IEEE Trans. Automat Control.*, **55** (5), 1047–1058.

20 Sarlette, A. and Sepulchre, R. (advisor) (2008) Geometry and symmetries in coordination control. PhD Thesis, University of Liège, Belgium.

21 Sarlette, A. and Sepulchre, R. (2009) Consensus optimization on manifolds. *SIAM J. Control and Optimization.* **48** (1), 56–76.

22 Sepulchre, R., Paley, D., and Leonard, N.E. (2004) Collective motion and oscillator synchronization, in *Cooperative Control*, vol. 309 (eds A. Morse, V. Kumar, and N. Leonard), Lecture Notes in Control and Information Science, Springer, pp. 189–205.

23 Jadbabaie, A., Motee, N., and Barahona, M. (2004) On the stability of the Kuramoto model of coupled nonlinear oscillators. Proceedings of the American Control Conference, June 30–July 2, 2004, Boston, MA, USA.

24 Canale, E. and Monzón, P. (2007) Gluing Kuramoto coupled oscillator networks. Proceedings of the 46th IEEE Conference on Decision and Control, December 12–14, 2007, New Orleans, LA, USA. pp. 4596–4601.

25 Sepulchre, R., Paley, D., and Leonard, N.E. (2007) Stabilization of planar collective motion with all-to-all communication. *IEEE Trans. Automat. Contr.*, **52** (5), 811–824.

26 Sarlette, A., Sepulchre, R., and Leonard, N.E. (2006) Discrete-time synchronization on the *N*-torus. Proceedings of the 17th International Symposium on Mathematical Theory of Networks and Systems, July 24–28, 2006, Kyoto, Japan. pp. 2408–2414.

27 Hopfield, J.J. (1982) Neural networks and physical systems with emergent collective computational capabilities. *Proc. Natl. Acad. Sci. USA*, **79**, 2554–2558.

28 Goles-Chacc, E., Fogelman-Soulie, F., and Pellegrin, D. (1985) Decreasing energy functions as a tool for studying threshold networks. *Discr. Appl. Math.*, **12** (3), 261–277.

29 Jeanne, J., Paley, D., and Leonard, N.E. (2005) On the stable phase relationships of loop-coupled planar particles.

Proceedings of the 44th IEEE Conference on Decision and Control, December 12–15, 2005, Sevilla, Spain.

30 Yeung, M.K.S. and Strogatz, S.H. (1999) Time delay in the Kuramoto model of coupled oscillators. *Phys. Rev. Lett.*, **82** (3), 648–651.

31 Boyd, S., Ghosh, A., Prabhakar, B., and Shah, D. (2006) Randomized gossip algorithms. *IEEE Trans. Inf. Theory (Special issue)*, **52** (6), 2508–2530.

32 Sarlette, A., Tuna, S.E., Blondel, V.D., and Sepulchre, R. (2008) Global synchronization on the circle. Proceedings of the 17th IFAC World Congress, July 6–11, 2008, Seoul, Korea.

33 Gallager, R.G., Humblet, P.A., and Spira, P.M. (1983) A distributed algorithm for minimum-weight spanning trees. *ACM Trans. Progr. Lang. Sys.*, **5** (1), 66–77.

34 Scardovi, L., Sarlette, A., and Sepulchre, R. (2007) Synchronization and balancing on the N-torus. *Sys. Control Lett.*, **56** (5), 335–341.

35 Chung, F.R.K. (1997) *Spectral Graph Theory*, vol. **92**, CBMS Regional Conference Series in Mathematics, AMS.

36 Diestel, R. (1997) *Graph Theory*, Springer.

Conclusion and Outlook

"The method of demonstration is not the same for the physicist as for the mathematician, but their methods of discovery are very similar. In the case of both they consist in rising from the fact to the law"... Henri Poincare (*Science and Method*, 1854–1912)

We would like to conclude this book by summarizing and discussing in a rather general context the role of dynamical systems and complexity in today's research. We will also give our opinion about what we believe will be the most important future developments in the field of complexity and dynamical systems.

The field of dynamical systems has evolved incredibly during the last decades and has turned out to be essential in understanding numerous phenomena, varying from feedback in laser systems, as discussed in Chapters 3 and 4, to turbulence in pipe flow addressed in Chapter 2, to name just a few. The combination of dynamical systems analysis with statistical physics has led to a thriving field that is now often referred to as complexity. Because statistical physics methods can also be applied to subjects such as social networks and epidemics, this merger has helped dynamical systems analysis to penetrate almost any research field in science. Because so many phenomena can be grouped under the umbrella of complexity, research in this field is flourishing and has resulted in numerous publications and new journals.

In this book, the broad spectrum of research that can be counted as belonging to dynamical systems/complexity is reflected. Part One of the book describes a large number of different examples (Chapters 1–6) that can be analyzed using the methods of either statistical physics or dynamical systems or a combination of the two. In some cases like those discussed in Chapter 6, both methods are applied to what turns out to be a stochastically driven dynamical system.

The results notwithstanding, we believe that dynamical systems and complexity will bring much more surprises in the future. The theory behind complex systems is still evolving. This is also covered in Part Two of the book. Emphasis in this endeavor is, for example, on symmetry. History has proved the effectiveness of looking at symmetries in many disciplines, especially in mathematics and physics. This approach is also valid for complex systems. For example, the absence (presence) of time symmetry may lead to the presence (absence) of an arrow of time (Chapter 7). Similarly, patterns caused by defects in lattices in solid-state physics, or even botany, can be connected to monodromy (Chapter 8). The choice of the sign of the monodromy is similar in spirit to time irreversibility. The more applied field of elasticity is another example where the combined statistical physics and dynamic systems approach has proved to be very fruitful (Chapter 9). The Boussinesq equation that arises from these considerations possesses strongly localized nonlinear solutions, usually called solitons. Systems of equations prone to pure solitonic type of dynamic behavior admit new conservation laws, which in the case of integrable systems correspond to symmetry properties. Solitons are still a very active research area and were encountered in applied areas such as pulsed laser systems and more remote areas such as string theory. In fact, soliton solutions were already suggested in the 1960s to resolve a famous problem that has long puzzled statistical physicists, the so-called Fermi–Pasta–Ulam problem. In the Fermi–Pasta–Ulam problem, a system consisting of N coupled weakly nonlinear oscillators were studied, using computer simulations. Surprisingly, it was found that when the initial configuration was such that the energy was initially concentrated in a single normal mode, the energy did not eventually distribute evenly among all oscillators as expected from the equipartition theorem of statistical physics. Zabusky and Kruskal gave an explanation by deriving a nonlinear partial differential equation that is equivalent to the Korteweg–de Vries equation and has soliton solutions. However, as solitonic solutions are related to weakly nonlinear partial differential equations that are integrable and nongeneric in the sense that any generic perturbation destroys the integrability, the explanation is still argued and alternative explanations such as KAM tori have been proposed. This story shows once more how statistical features and dynamics are intertwined. For resolving certain classes of problems, both disciplines are essential.

Finally, we should mention and emphasize the importance of networks. This field has experienced a giantic impetus with the work of Barabasi, Newman, and Albert, among others. Many disciplines have embraced the theory of complex networks, ranging from social siences, epidemiology, neuronal networks, and predictions of forest fires and earthquakes, while it has also been used in explaining power failures in power grids and the celebrated page rank algorithm used by Google to order the web sites on the Internet. The network topology plays a crucial role in most problems related to networks, and still many problems concerning network topologies and the way the dynamics is influenced by its underlying topology remain unresolved; see, for example, Chapter 10 in which for a circular topology the synchronization problem is examined. An interesting question is how network topology evolves from certain connection probabilities. These questions are particularly important for neuronal networks and are still unsolved. Particular features such as the small world or scale-free network property can occur only for certain kinds of dynamical interactions between vertices in the network. Finding a correspondence between vertex interactions and network topology is still an unresolved problem.

We believe that the road ahead for complex dynamical systems is very promising. Besides the theoretical challenges that remain, more and more applications appear to fit well in the scheme of complexity and are amenable to more reliable methods that are becoming available to analyze these systems.

Index

a

action-angle variables   160, 162
agents   213–239
angular momentum   166
anomalous diffusion   121, 124, 129, 131, 134
anomalous dispersion   194
anticonsensus configuration   226
arrow of time   141, 142, 148, 154
autocorrelation function   79, 80

b

balancing configuration   226, 234
bed perturbations   21, 23
Bessel function   42–44, 89
bidromy   174
bifurcation parameter   40, 81, 82, 85, 86
biological systems   99–101, 105, 116, 137
Blausius–Huppert–Stone (BHS) model   111
Bogdanov–Takens point. *see* double zero eigenvalues
Boussinesq equation   192, 194
– generalized   195
Boussinesq paradigm   191
Brouwer's fixed point theorem   104

c

Cauchy-stress   187
chaos   63, 83, 84, 87, 90–95, 111, 139
circadian rhythm   107, 108, 113, 114
cis side   119, 123, 125, 128
consensus   214–220, 225, 229, 234
continuous time random walk (CTRW)   124, 125
coupling constant   44
covariant vector   188
crescentic bar   9, 10, 17–24
cusp point   51, 53, 54, 57

d

d' Alembert's equation   184
delay differential equation   38
density matrix   153
depth-averaged concentration   12, 13
depth-averaged velocity   10, 11
detailed balance   125, 142–144, 147
detuning   47, 71, 72, 75
directed graphs   219, 227, 231
dispersion   184, 191, 194, 195, 197, 200, 205, 209
dissipated work   151, 152
DNA   109, 119, 125, 137
– packing   109
double zero eigenvalues   47, 48, 51
dynamic materials   184, 207

e

effective Hamiltonian   159
eigenfunction expansion   39, 43
eigenvalues   20, 40, 159, 164, 165, 237
– complex   15, 55
– double zero   47, 48, 51
elementary cell   176–178
elementary singularities   171
energy-momentum map   160, 162–167, 173–175
entropic exponent   122
entropy   82, 141–156, 203
– Kolmogorov–Sinai   100
– microscopic   155, 156
– production   141–156
equilibrium   14, 15, 17, 21–23, 121–123, 125, 141–147, 151–156, 183, 223, 225, 226, 230, 231
Eshelby stress   186, 191, 203
Euler equations   10, 190

The Complexity of Dynamical Systems. Edited by J. Dubbeldam, K. Green, and D. Lenstra
Copyright © 2011 WILEY-VCH Verlag GmbH & Co. KGaA, Weinheim
ISBN: 978-3-527-40931-0

extensive quantity   154
external cavity feedback
external cavity modes   39, 40, 59
external cavity oscillations   51, 56–60

f
feedback   99
– cross   39, 45, 46, 48, 56–60
– delayed   37
– negative   101–107, 109–117
– optical   40, 42
– positive   109
– self   39, 45–47, 59
FENE potential   127
first passage time distribution   126–129, 131–134
fluctuation theorems   149, 150, 154
Fokker–Planck equation   120–122, 124, 127, 135, 150
fold points. see saddle node bifurcation
fractional diffusion equation (FDE)   125, 129
fractional Fokker–Planck equation (FFPE)   124, 125, 129, 130, 135, 136
Frankel–Kontorova model   197
frequency filtering transfer function   79
Froude numbers   17
frustration   104, 114, 115

g
generalized continua   184
generic singularity   178
genes   108, 113, 114
gossip algorithm   215, 231–234
gradient algorithms   222, 234
grain dynamics   24
graph theory   235
Green–Kubo expressions   142
group, orbit   41, 46
– fundamental   162
– homology   162

h
Hagen–Poiseuille flow   27
Hastings–Powell (HP) model   110
HCN molecule   165
homoclinic bifurcation   56, 57, 59
homotopy   161
homotopy parameter   45, 47, 52, 57, 59
Hopf bifurcation   41, 42, 48–57, 68, 82, 89, 103, 113, 115, 163
Hugoniot conditions   200, 201, 204

i
Ikeda dynamics   82
Ikeda map   82
Ikeda ring cavity   64, 65, 67, 81.
integrable systems   160, 163, 164, 173, 175
irreversibility   142, 146, 149, 155, 178, 200

j
Jacobi–Anger expansion   89

k
Kawahara solitons   195
Kerr effect   66, 81
Klein bottle   172
Korteweg-de Vries equation   184, 192
Kuramoto model   213, 214, 220

l
Lagrangian   185, 187, 190, 193, 210
Landauer principle   155
Lang–Kobayashi equations   38, 40, 68, 72
Langevin equation   93, 150
large scale sediment   25
Levy flight   124
LIDAR   95
limit cycle   82, 89, 111, 214, 224, 226
linear filtering   64, 66, 81
Liouville's theorem   143, 152
long-lived transients   27–29
long-wavelength limit   201
low-frequency fluctuations (LFF)   69, 71

m
Mackey–Glass dynamics   82, 83
Manakov top   173
Markov chain   233
Markovian description   144, 150
Markovian dynamics   100
Markovian master equation   147
Markovianity   148
material manifold   203
Mdm2   106, 107
mean first passage time   126
Metropolis algorithm   128
Mittag–Leffler function   126–128, 130
– generalized   131, 132
monodromy   159–161, 164–167
– defects   168, 170, 177
– fractional   163, 169
– quantum   161, 164, 165, 170
– transformation   162, 173
monotonicity   98, 102, 104
Monte Carlo simulations   121, 122, 127, 133

morphodynamics 7–14
Morse potential 128

n

Navier–Stokes equations 27
negative feedback loop (NFL) 101–107
network 95, 109
– communication 37
– Hopfield 224
– optic 91
– oscillators 78
– transition 109, 111, 113, 116
noise 85, 92–94, 105, 119, 214
nonextensive systems 141, 142
nonlinear wave equation 189, 192
nonlocality 184, 191, 201, 204
nullclines 98, 102, 104
numerical bifurcation analysis 40

o

OCCULT project 90
Onsager relations 142, 144
oscillators 63, 220
– delay-coupled 73, 79
– weakly coupled 213
oscillatory dynamics 101

p

period doubling 56, 57, 82, 89, 111
phase interference condition 67
phase transition 184, 195, 196, 200, 201, 203–207, 213
phase variables 213, 220
phase-averaged morphodynamics 17
phase-averaged velocity 18
phonons 184, 188
Piola–Kirchoff stress 186, 203
Pockels effect 67
Poisson bracket 160
polarized optical modes 42, 43
puff 27, 29–36
pump current 38–40, 44, 75

q

quality factor (Q) 88, 93
quantum numbers 159, 164, 167
quantum systems 159, 160
quasi particles 184, 188, 193, 198, 200, 205
quasi-equilibrium 135–137
quasi-periodic motion 227

r

randomness 119, 231
Rankine–Hugoniot equation 204

rate equations 40, 44
regulatory systems 99
relaxation oscillation (RO) 42, 51, 58, 60, 65, 66
repressor 100–115
Reynolds number 27–35
rheonomic materials 187
Riemann–Liouville derivative 125
RNA polymerase 109

s

saddle node bifurcation 41, 49–57
scaling 15, 32–35, 82, 120, 128
– arguments 132, 133
scleronomic materials 187, 189
second law of thermodynamics 141, 142, 155, 156, 202, 203
sector 99–105
sediment movement 25
sediment transport 9–12, 24
self-coupling 45, 69
self-rotation 177, 178
semiconductor laser 37, 38, 57, 64, 65, 67–69, 73, 78, 92, 95
shear strain 195, 202
sine-Gordon equation 184, 197, 199
singular fibers 163, 166
slug 36
solitary waves 184, 190, 192
solitons 190, 192–196, 198
spiral phyllotaxis 161, 175, 176
statistical moments 132, 135
stochastic thermodynamics 142, 144, 152
– pathwise 144
strain 153, 184, 185, 187, 190, 194, 196, 201–204
strange attractor 29, 35
surface gravity waves 7, 16, 25
swarms 213–218, 224, 226, 229–231
symbolic dynamics 99–113
symmetry breaking 69, 72, 115, 203
symmetry relation 203
symmetry, S^1 56
– π 56
– SL(2Z) 163
– S^n, SO(n) 215, 234
– translational 41, 42, 203
synchronization 63, 66, 75, 77, 90, 95, 213, 214
– generalized 71, 72, 74
– on a circle 213–238
– zero-lag 73–75, 77

t

thermoelasticity 206
topological invariant 177

tori 174
toric fibrations 160, 161, 173–175, 174
toric foliations 173
trans side 119, 121, 128, 136
transition diagram 101
translational symmetry 41, 42, 203
translocation 119
– driven 130
– time 119, 128
– unbiassed 122
travelling waves 28
tunneling 165
turbulence 22, 27, 29, 33, 35

u
undirected graphs 222, 226, 229, 236
unfolding 173, 174

v
VCSEL 38–47, 59
Vicsek model 213–215, 219, 220, 226, 229, 235

w
wave speed 9, 13–17, 188, 190
weakly nonlinear methods 24
Wiener–Khinchine theorem 79
wind waves 7